Titles in This Series

W0037735

Volume

1 **Markov random fields and their applications**, Ross Kindermann and J. Laurie Snell

2 **Proceedings of the conference on integration, topology, and geometry in linear spaces**, William H. Graves, Editor

3 **The closed graph and P-closed graph properties in general topology**, T. R. Hamlett and L. L. Herrington

4 **Problems of elastic stability and vibrations**, Vadim Komkov, Editor

5 **Rational constructions of modules for simple Lie algebras**, George B. Seligman

6 **Umbral calculus and Hopf algebras**, Robert Morris, Editor

7 **Complex contour integral representation of cardinal spline functions**, Walter Schempp

8 **Ordered fields and real algebraic geometry**, D. W. Dubois and T. Recio, Editors

9 **Papers in algebra, analysis and statistics**, R. Lidl, Editor

10 **Operator algebras and K-theory**, Ronald G. Douglas and Claude Schochet, Editors

11 **Plane ellipticity and related problems**, Robert P. Gilbert, Editor

12 **Symposium on algebraic topology in honor of José Adem**, Samuel Gitler, Editor

13 **Algebraists' homage: Papers in ring theory and related topics**, S. A. Amitsur, D. J. Saltman, and G. B. Seligman, Editors

14 **Lectures on Nielsen fixed point theory**, Boju Jiang

15 **Advanced analytic number theory. Part I: Ramification theoretic methods**, Carlos J. Moreno

16 **Complex representations of GL(2, K) for finite fields K**, Ilya Piatetski-Shapiro

17 **Nonlinear partial differential equations**, Joel A. Smoller, Editor

18 **Fixed points and nonexpansive mappings**, Robert C. Sine, Editor

19 **Proceedings of the Northwestern homotopy theory conference**, Haynes R. Miller and Stewart B. Priddy, Editors

20 **Low dimensional topology**, Samuel J. Lomonaco, Jr., Editor

21 **Topological methods in nonlinear functional analysis**, S. P. Singh, S. Thomeier, and B. Watson, Editors

22 **Factorizations of $b^n \pm 1$, $b = 2, 3, 5, 6, 7, 10, 11, 12$ up to high powers**, John Brillhart, D. H. Lehmer, J. L. Selfridge, Bryant Tuckerman, and S. S. Wagstaff, Jr.

23 **Chapter 9 of Ramanujan's second notebook—Infinite series identities, transformations, and evaluations**, Bruce C. Berndt and Padmini T. Joshi

24 **Central extensions, Galois groups, and ideal class groups of number fields**, A. Fröhlich

25 **Value distribution theory and its applications**, Chung-Chun Yang, Editor

26 **Conference in modern analysis and probability**, Richard Beals, Anatole Beck, Alexandra Bellow, and Arshag Hajian, Editors

27 **Microlocal analysis**, M. Salah Baouendi, Richard Beals, and Linda Preiss Rothschild, Editors

28 **Fluids and plasmas: geometry and dynamics**, Jerrold E. Marsden, Editor

29 **Automated theorem proving**, W. W. Bledsoe and Donald Loveland, Editors

30 **Mathematical applications of category theory**, J. W. Gray, Editor

31 **Axiomatic set theory**, James E. Baumgartner, Donald A. Martin, and Saharon Shelah, Editors

32 **Proceedings of the conference on Banach algebras and several complex variables**, F. Greenleaf and D. Gulick, Editors

Titles in This Series

Volume

33 **Contributions to group theory,** Kenneth I. Appel, John G. Ratcliffe, and Paul E. Schupp, Editors

34 **Combinatorics and algebra,** Curtis Greene, Editor

35 **Four-manifold theory,** Cameron Gordon and Robion Kirby, Editors

36 **Group actions on manifolds,** Reinhard Schultz, Editor

37 **Conference on algebraic topology in honor of Peter Hilton,** Renzo Piccinini and Denis Sjerve, Editors

38 **Topics in complex analysis,** Dorothy Browne Shaffer, Editor

39 **Errett Bishop: Reflections on him and his research,** Murray Rosenblatt, Editor

40 **Integral bases for affine Lie algebras and their universal enveloping algebras,** David Mitzman

41 **Particle systems, random media and large deviations,** Richard Durrett, Editor

42 **Classical real analysis,** Daniel Waterman, Editor

43 **Group actions on rings,** Susan Montgomery, Editor

44 **Combinatorial methods in topology and algebraic geometry,** John R. Harper and Richard Mandelbaum, Editors

45 **Finite groups–coming of age,** John McKay, Editor

46 **Structure of the standard modules for the affine Lie algebra $A_1^{(1)}$,** James Lepowsky and Mirko Primc

47 **Linear algebra and its role in systems theory,** Richard A. Brualdi, David H. Carlson, Biswa Nath Datta, Charles R. Johnson, and Robert J. Plemmons, Editors

48 **Analytic functions of one complex variable,** Chung-chun Yang and Chi-tai Chuang, Editors

49 **Complex differential geometry and nonlinear differential equations,** Yum-Tong Siu, Editor

50 **Random matrices and their applications,** Joel E. Cohen, Harry Kesten, and Charles M. Newman, Editors

51 **Nonlinear problems in geometry,** Dennis M. DeTurck, Editor

52 **Geometry of normed linear spaces,** R. G. Bartle, N. T. Peck, A. L. Peressini, and J. J. Uhl, Editors

53 **The Selberg trace formula and related topics,** Dennis A. Hejhal, Peter Sarnak, and Audrey Anne Terras, Editors

54 **Differential analysis and infinite dimensional spaces,** Kondagunta Sundaresan and Srinivasa Swaminathan, Editors

55 **Applications of algebraic K-theory to algebraic geometry and number theory,** Spencer J. Bloch, R. Keith Dennis, Eric M. Friedlander, and Michael R. Stein, Editors

56 **Multiparameter bifurcation theory,** Martin Golubitsky and John Guckenheimer, Editors

57 **Combinatorics and ordered sets,** Ivan Rival, Editor

58 **The Lefschetz centennial conference. Part I: Proceedings on algebraic geometry,** D. Sundararaman, Editor

The Lefschetz Centennial Conference

Part I: Proceedings on Algebraic Geometry

SOLOMON LEFSCHETZ

CONTEMPORARY
MATHEMATICS

Volume 58

The Lefschetz
Centennial Conference

Part I: Proceedings on
Algebraic Geometry

Proceedings of The Lefschetz
Centennial Conference held
December 10–14, 1984

D. Sundararaman, Editor

AMERICAN MATHEMATICAL SOCIETY
Providence • Rhode Island

The Proceedings of The Lefschetz Centennial Conference on Algebraic Geometry, Algebraic Topology and Differential Equations was held at the Centro de Investigación y de Estudios Avanzados, in Mexico City, Mexico December 10–14, 1984.

1980 *Mathematics Subject Classification*. Primary 14-06.

Organizing Committee: J. Adem, S. Gitler, J. J. Kohn, E. Ramirez de Arellano, H. Rossi, D. Sundararaman, A. Verjovsky.

Editorial Committee of the Proceedings: D. Sundararaman, J. Adem, S. Gitler, E. Ramirez de Arellano, A. Verjovsky.

Library of Congress Cataloging-in-Publication Data

Lefschetz Centennial Conference (1984 : Mexico City, Mexico)
The Lefschetz Centennial Conference.
(Contemporary mathematics, ISSN 0271-4132; v. 58)
Bibliography: v. 1, p.
Contents: pt. 1. Proceedings on algebraic geometry.
1. Algebraic geometry—Congresses. 2. Algebraic topology—Congresses. 3. Differential equations—Congresses. 4. Lefschetz, Solomon, 1884–1972. I. Sundararaman, D. II. Series: Contemporary mathematics (American Mathematical Society); v. 58.
QA564.L43 1984 512′.33 86-14040
ISBN 0-8218-5065-2 (set: alk. paper)
ISBN 0-8218-5061-X (v. 1: alk. paper)

Printed in the United States of America.
This volume was printed directly from author-prepared copy.
The paper used in this book is acid-free and falls within the guidelines established to ensure permanence and durability.

TABLE OF CONTENTS

The following lectures do not appear in this volume:

An overview of the Schottky problem
E. ARBARELLO

Fermi curves and monodromy
D. GIESEKER

Local isometric embeddings of Riemannian manifolds
P. GRIFFITHS

Some results on the equation $x^3 + y^3 = Az^3$
Z. RAMOS

PREFACE

The Solomon Lefschetz Centennial Conference was held in the Centro de Investigación y de Estudios Avanzados in Mexico City from 10 to 14, December 1984. The topics of the conference were Algebraic Geometry, Algebraic Topology and Differential Equations, the three areas to which the Late S. Lefschetz had made significant contributions. Besides the mathematicians in Mexico, experts from many countries participated actively and enthusiastically resulting in the very successful conference.

This volume contains the proceedings of the conference corresponding to the general area of Algebraic Geometry. It also contains Solomon Lefschetz' article "A Page of Mathematical Autobiography", which is reproduced, with permission, from the Bulletin of the American Mathematical Society and William Hodge's article "Solomon Lefschetz", which is reproduced, with permission, from the Bulletin of the London Mathematical Society.

Financial support to the conference came from CONACyT and SEP of Mexico, and the NSF of U.S.A. cosponsored and supported the conference through their Cooperative Science Programs with Latin America and the Caribbean.

The Organizing Committee of the Conference and the Editorial Committee of the Proceedings would like to thank all the participants in the conference and to all the contributors to this volume.

D. Sundararaman.

CINVESTAV, Mexico City, April 1986.

Contemporary Mathematics
Volume 58, Part I, 1986

A PAGE OF MATHEMATICAL AUTOBIOGRAPHY

BY SOLOMON LEFSCHETZ

INTRODUCTION

As my natural taste has always been to look forward rather than backward this is a task which I did not care to undertake. Now, however, I feel most grateful to my friend Mauricio Peixoto for having coaxed me into accepting it. For it has provided me with my first opportunity to cast an objective glance at my early mathematical work, my algebro-geometric phase. As I see it at last it was my lot to plant the harpoon of algebraic topology into the body of the whale of algebraic geometry. But I must not push the metaphor too far.

The time which I mean to cover runs from 1911 to 1924, from my doctorate to my research on fixed points. At the time I was on the faculties of the Universities of Nebraska (two years) and Kansas (eleven years). As was the case for almost all our scientists of that day my mathematical isolation was complete. This circumstance was most valuable in that it enabled me to develop my ideas in complete mathematical calm. Thus I made use most uncritically of early topology à la Poincaré, and even of my own later developments. Fortunately someone at the Académie des Sciences (I always suspected Émile Picard) seems to have discerned "the harpoon for the whale" with pleasant enough consequences for me.

To close personal recollections, let me tell you what made me turn with all possible vigor to topology. From the ρ_0 formula of Picard, applied to a hyperelliptic surface Φ (topologically the product of 4 circles) I had come to believe that the second Betti number $R_2(\Phi)=5$, whereas clearly $R_2(\Phi)=6$. What was wrong? After considerable time it dawned upon me that Picard only dealt with *finite* 2-cycles, the only useful cycles for calculating periods of certain double integrals. Missing link? The cycle at infinity, that is the plane section of the surface at infinity. This drew my attention to cycles carried by an algebraic curve, that is to *algebraic* cycles, and · · · the harpoon was in!

An address delivered at Brown University on April 14, 1967. Submitted by invitation of the editors; received by the editors September 7, 1967.

My general plan is to present the first concepts of algebraic geometry, then follow up with the early algebraic topology of Poincaré plus some of my own results on intersections of cycles. I will then discuss the topology of an algebraic surface. The next step will be a summary presentation of the analytical contributions of Picard, Severi and Poincaré leading to my work, application of topology to complex algebraic geometry concluding with a rapid consideration of the effect on the theory of abelian varieties.

This is not however a cold recital of results achieved duly modernized. To do this would be to lose the "autobiographical flavor" of my tale. I have therefore endeavored to place myself back in time to the period described and to describe everything as if I were telling it a half century ago. From the point of view of rigor there is no real loss. Analytically the story is fairly satisfactory and to make it so in the topology all that is needed is to accept the results amply described in my Colloquium Lectures [**10**].

To place the story into focus I must say something about what we knew and accepted in days gone by. That is I must describe our early background.

In its early phase (Abel, Riemann, Weierstrass), algebraic geometry was just a chapter in analytic function theory. The later development in this direction will be fully described in the following chapters. A new current appeared however (1870) under the powerful influence of Max Noether who really put "geometry" and more "birational geometry" into algebraic geometry. In the classical mémoire of Brill-Noether (Math. Ann., 1874), the foundations of "geometry on an algebraic curve" were laid down centered upon the study of linear series cut out by linear systems of curves upon a fixed curve $f(x, y) = 0$. This produced birational invariance (for example of the genus p) by essentially algebraic methods.

The next step in the same direction was taken by Castelnuovo (1892) and Enriques (1893). They applied analogous methods to the creation of an entirely new theory of algebraic surfaces. Their basic instrument was the study of linear systems of curves on a surface. Many new birationally invariant properties were discovered and an entirely new and beautiful chapter of geometry was opened. In 1902 the Castelnuovo-Enriques team was enriched by the brilliant personality of Severi. More than his associates he was interested in the contacts with the analytic theory developed since 1882 by Émile Picard. The most important contribution of Severi, his theory of the base (see §12) was in fact obtained by utilizing the Picard number ρ (see §11).

The theory of the great Italian geometers was essentially, like Noether's, of algebraic nature. Curiously enough this holds in good part regarding the work of Picard. This was natural since in his time Poincaré's creation of algebraic topology was in its infancy. Indeed when I arrived on the scene (1915) it was hardly further along.

About 1923 I turned my attention to "fixed points" which took me away from algebraic geometry and into the more rarefied air of topology. I cannot therefore refer even remotely to more recent doings in algebraic geometry. I cannot refrain, however, from mention of the following noteworthy activities:

I. The very significant work of W. V. D. Hodge. I refer more particularly to his remarkable proof that an n-form of V^q which is of the first kind cannot have all periods zero (see Hodge [**13**]).

II. The systematic algebraic attack on algebraic geometry by Oscar Zariski and his school, and beyond that of André Weil and Grothendieck. I do feel however that while we wrote algebraic GEOMETRY they make it ALGEBRAIC geometry with all that it implies.

References. For a considerable time my major reference was the Picard-Simart treatise [**2**]. In general however except for the writings of Poincaré on topology my Borel series monograph [**9**] is a central reference. The best all around reference not only to the topics of this report but to closely related material is the excellent Ergebnisse monograph of Zariski [**11**]. Its bibliography is so comprehensive that I have found it unnecessary to provide an extensive one of my own.

TABLE OF CONTENTS

I. General Remarks on Algebraic Varieties

1. **Definition. Function field.** It was the general implicit or explicit understanding among algebraic geometers of my day that an algebraic n-variety V^n (n dimensional variety) is the partial or complete irreducible intersection of several complex polynomials or "hypersurfaces" of a projective space S^{n+k}, in which V^n had no singularities (it was homogeneous). Thus V^n was a compact real $2n$-manifold M^{2n} (complex dimension n). It could therefore be considered as its own Riemann manifold as I shall do throughout.

For convenience in analytical operations one customarily represents V^n by a general projection in cartesian S^{n+1}

$$F(x_1, x_2, \cdots, x_n, y) = 0, \tag{1.1}$$

where F is an irreducible complex polynomial of degree m. In this representation, the variety, now called F, occupies no special position relative to the axes.[1] As a consequence (1.1) possesses the simplest singularities. For a curve they consist of double points with distinct tangents, for a surface: double curve with generally distinct tangent planes along this curve.

Incidentally, the recent brilliant reduction of singularities by Hironaka [12] has shown that the varieties as just described are really entirely general.

Returning to our V^n the study of its topology will lean heavily upon the properties of the pencil of hypersurfaces $\{H_y\}$ cut out by the hyperplanes $y = \text{const}$. The particular element of the pencil cut out by $y = c$ is written H_c. As my discourse will be mostly on surfaces I will only describe (later) certain pecularities for varieties.

Function field. Let the complex rational functions $R(x_1, \cdots, x_n, y)$ be identified mod F. As a consequence they constitute an algebraic extension of the complex field K written $K(F)$, called the function field of F.

Let F^* be the nonsingular predecessor of F in S^{n+k} and let $(u_1, \cdots, u_{n+k})$ be cartesian coordinates for S^{n+k}. On F^* they determine elements ξ_h, $h \leqq n+k$ of $K(F)$. The system

$$u_h = \xi_h, \qquad h \leqq n + k$$

is a parametric representation of F^*. F^* is a *model* of $K(F)$.

Any two models F_1^*, F_2^* are birationally equivalent: birationally transformable into one another. The properties that will mainly

[1] That is, F has only those singularities which arise from a general projection on S^{n+1} of a nonsingular $V^n \subset S^{n+k}$.

interest us are those possessing a certain degree of birational invariance (details in §17).

Terminology. Since only algebraic curves, surfaces, varieties will be dealt with, I drop the mention "algebraic" and merely say curve, ...

The symbol $\mathcal{V}^n$ represents a (usually complex) n dimensional vector space.

2. **Differential forms.** Let $\alpha, \beta, \cdots$, denote elements of the function field $K(F)$. I shall refer to various differentials: zero, one, two, $\cdots$ forms $\omega^0, \omega^1, \omega^2, \cdots$, in the sense of Élie Cartan of type

$$\omega^k = \sum \alpha_{i_1,\ldots,i_k} d\alpha_{i_1} \cdots d\alpha_{i_k},$$

every α in $K(F)$, as zero, one, two, $\cdots$, forms. They are calculated by the rules of calculus, remembering that the $d\alpha_j$ are skew-symmetric, that is $d\beta d\alpha = -d\alpha d\beta$.

Note that $d\omega^k$ is an ω^{k+1} called *exact* and that if $d\omega^k = 0$ one says that ω^k is *closed.*

Special terms are: ω^k is of the first kind when it is holomorphic everywhere on F; of the second kind when it is holomorphic at any point of F mod some $d\alpha$; of the third kind if neither of the first nor of the second kind.

The evaluation of the number of kinds one or two constitutes one of the main problems to be discussed.

3. **Differential forms on a curve.** Let the curve be

$$f(x, y) = 0 \tag{3.1}$$

and let m be its degree. We refer to it as "the curve f." Under our convention, f has no other singularities than double points with distinct tangents and is identified in a well-known sense with its Riemann surface. Its one-forms are said to be *abelian.* An *adjoint* to f is a polynomial $\phi_n(x, y)$ (n is its degree) vanishing at all double points.

The following are classical properties:

One-forms of the first kind. They are all reducible to the type

$$\frac{\phi_{m-3}dx}{f'_y}. \tag{3.2}$$

They form a $\mathcal{V}^p$, where $2p = R_1$, the first Betti number of the Riemann surface f. Of course the collection $\{\phi_{m-3}\}$ forms likewise a $\mathcal{V}^p$.

One-forms of the second kind. Same type of reduction to (3.2) mod a $d\omega^0$, save that ϕ_{m-3} is replaced by some ϕ_s. Their vector space mod $dK(f)$ is a $\mathcal{V}^{2p}$.

One-forms of the third kind. They have a finite number of logarithmic points with residues whose sum is zero.

Some special properties of one-forms of the first kind. Let

$$\psi = \sum_{h=0}^{r} \alpha_k \psi_h(x, y) = 0 \tag{3.3}$$

be a linear system of polynomials linearly independent mod f and of common degree. Let the general ψ intersect f in a set of points $P_1, \cdots, P_s$ which includes all the variable points and perhaps some fixed points. The collection of all such sets is a *linear series of degree* n and *dimension* r. The series is *complete* when its sets do not belong to an amplified series of the same degree: designation g_n^r (concepts and terminology of Brill and Noether).

(3.4) THEOREM OF ABEL. *Let du be any one-form of the first kind; let* $\{P_h\}$ *be any element of a* g_n^r *and let A be a fixed point of f. Then with integration along paths on f:*

$$\sum \int_A^{P_h} du = v$$

is a constant independent of the element $\{P_h\}$ *of* g_n^r.

Still another classic, a sort of inverse of Abel's theorem is this:

(3.5) THEOREM OF JACOBI. *Let* $\{du_h\}$ *be a base for the one-forms of the first kind. Then for general values of the constants* v_h *(exceptions noted) the system*

$$\sum_{k=1}^{p} \int_A^{P_k} du_h = v_h$$

in the p unknowns P_k, $k \leqq p$, *has a unique solution.*

Periodic properties. Let $\{du_h\}$ be as just stated and let $\{\gamma_\mu^1\}$, $\mu \leqq 2p$ be an integral homology base (see (5.4)) for the module of one-cycles of f. The expression

$$\pi_{h\mu} = \int_{\gamma_\mu^1} du_h$$

is the *period* of $\int du_h$ as to the cycle γ_μ^1. Let the matrix

$$\Pi = [\pi_{h\mu}]; \qquad h, \mu \leqq 2p; \qquad \pi_{h+p,\mu} = \bar{\pi}_{h\mu}, \; h \leqq p.$$

By means of integration on the Riemann surface f, Riemann has obtained the following comprehensive result (formulation of Scorza):

(3.6) THEOREM OF RIEMANN. *There exists an integral skew-symmetric $2p \times 2p$ matrix M with invariant factors unity such that*

$$(3.7) \qquad i\Pi M \Pi' = \begin{bmatrix} 0 & A \\ A^* & 0 \end{bmatrix}, \qquad (A^* = \bar{A}')$$

is a positive definite hermitian matrix.

Riemann matrices. This is the name given by Scorza to a matrix like Π satisfying a relation (3.7) except that M is merely rational skew-symmetric. The theory of such matrices has been extensively developed by Scorza [6]. He called M: *principal matrix* of Π.

It may very well happen that there is more than one rational skew-symmetric matrix M satisfying a relation (3.7) but without necessarily the positive definite property. These matrices are called singularity matrices. They form a rational vector space whose dimension k is the *singularity index* of the Riemann matrix (Scorza).

II. TOPOLOGY

4. **Results of Poincaré.** Let M^n be a compact orientable n-manifold which admits a cellular subdivision with α_k k-cells (well-known property for varieties). The characteristic is the expression

$$(4.1) \qquad \chi(M^n) = \sum (-1)^k \alpha_k.$$

The following two relations were proved by Poincaré:

$$(4.2) \qquad \chi(M^n) = \sum (-1)^k R_k$$

$$(4.3) \qquad R_k = R_{n-k}$$

where R_k is the kth integral Betti number of M^n: maximum number of linearly independent k-cycles with respect to homology (=with respect to bounding).

5. **Intersections.** In my work on algebraic geometry I freely used the intersection properties described below; they were actually justified and proved topologically invariant a couple of years later in my paper in the 1926 Transactions and much more fully in [**10**].

Let M^n be as before and let γ^p and γ^q be integral p- and q-cycles of M^n. One may define the intersection $\gamma^p \cdot \gamma^q$ and it is a $(p+q-n)$-cycle.

(5.1) *If γ^p or $\gamma^q \sim 0$ (bounds), then also $\gamma^p \cdot \gamma^q \sim 0$.*

The more important situation arises when $p+q=n$. The intersection (geometric approximation) is then a zero-cycle

$$C^0 = \sum s_j A_j$$

where the s_j are integers. The *intersection number*

$$(\gamma^p, \gamma^{n-p}) = \sum s_j$$

is independent of the approximation. One proves readily

$$(5.2) \qquad (\gamma^p, \gamma^{n-p}) = (-1)^{(n-p)p}(\gamma^{n-p}, \gamma^p).$$

A basic result is:

(5.3) THEOREM. *A n.a.s.c. in order that* $\lambda\gamma^p \sim 0$, $\lambda \neq 0$ *is that*

$$(\gamma^p, \gamma^{n-p}) = 0$$

for every γ^{n-p} [9, p. 15], [10, p. 78].

(5.4) HOMOLOGY BASE. The collection $\{\gamma_h^p\}$, $h \leqq R_p$ is a homology base for the p-cycles when the γ_h^p are independent and every γ^p satisfies a relation

$$\lambda\gamma^p \sim \sum s_h \gamma_h^p, \qquad \lambda \neq 0.$$

(5.5) *A n.a.s.c. in order that the* $\{\gamma_h^p\}$, $h \leqq R_p$ *be a homology base for* p*-cycles is the existence of a set of* R_p *cycles* $\{\gamma_k^{n-p}\}$ *such that the determinant*

$$|(\gamma_h^p, \gamma_k^{n-p})| \neq 0.$$

Then $\{\gamma_k^{n-p}\}$ *is likewise a homology base for* $(n-p)$*-cycles.*

6. **The surface** F. **Orientation.** Let P be a point of F and let $u = u' + iu''$, $v = v' + iv''$ be local coordinates for P. Orient F by naming the real coordinates in the order u', u'', v', v''. There results a unique and consistent orientation throughout the surface F. Hence F is an orientable M^4.

Similarly if C is a curve of F and u is a local coordinate at a nonsingular point Q of C. The resulting orientation turns C into a definite two-cycle, still written C.

Let D be a second curve through Q, for which Q is nonsingular and not a point of contact of the two curves. Then Q contributes $+1$ to both the intersection number (C, D) and to the number $[CD]$ of geometric intersections of C and D. This holds also, through certain approximations when Q is a multiple intersection. Hence always

$$(6.1) \qquad (C, D) = [CD].$$

I will return to these questions later.

7. **Certain properties of the surface F. Its characteristic.** To be a little precise let for a moment F^* denote the nonsingular predecessor of F in projective S^{k+2}. One may always choose a model F^* of the function field $K(F)$ whose hyperplane sections are in general of a fixed genus $p>0$. We pass now to a cartesian representation of degree m:

$$(7.1) \qquad F(x, y, z) = 0$$

which is a general projection of F^* and in particular in general position relative to the axes. The general scheme that follows is due to Picard. Let $\{H_y\}$ be the pencil cut out by the planes $y=\text{const.}$, and let a_h, $h \leqq N$, be the values for which the planes $y=a_h$ are tangent to F. Then the following properties hold:

I. Every H_y, y not an a_k, is of fixed genus p.

II. Every H_y is irreducible.

III. The plane $y=a_k$ has a unique point of contact A_k with F and A_k is a double point of H_{a_k} with distinct tangents. Hence the genus of H_{a_k} is $p-1$.

IV. Among the branch points of the function $z(x)$ taken on H_y exactly two $\to A_k$ as $y \to a_k$.

V. The fixed points $P_1, \cdots, P_m$ of H_y are all distinct.

I denote by S_y the sphere of the complex variable y.

Characteristic. Cover H_y with a cellular decomposition among whose vertices are the fixed points P_h of the curve.

Then if $H_y^* = H_y - \sum P_h$, $\chi(H_y^*) = 2-2p-m$. Decompose also S_y into cells with the a_k as vertices. Were it not for these points, and since a sphere has characteristic two, H_y^* promenading over S_y would generate a set $E = S_y \times H_y^*$ of characteristic

$$\chi(E^*) = 2(2 - 2p - m).$$

Now in comparison with H_y^*, H_{a_k} has lost two one-cycles, and has two points replaced by one. Hence

$$\chi(\overset{*}{H}_{a_k}) = \chi(\overset{*}{H}_y) + 1.$$

Upon remembering to add the missing points P_h we have then

$$(7.2) \quad \chi(F) = \chi(E^*) + N + m = (N - m - 4p) + 4 = I + 4$$

a formula due to J. W. Alexander (different proof). The number $I = N - m - 4p$ is the well-known *invariant of Zeuthen-Segre.*

8. **One-cycles of** F. The first step was taken by Picard who proved this noteworthy result:

(8.1) THEOREM. *Every one-cycle* γ^1 *of* F *is* $\sim$ *a cycle* γ^1 *contained in an* H_y.

The next important observation made by Picard was that H_y contained a certain number r of one-cycles which are invariant as y varies. That is such a cycle γ^1 situated say in H_a (a not an a_k) has the property that as y describes any closed path from a to a on the sphere S_y the cycle γ^1 returns to a position $\gamma^1 \sim \gamma^1$ in H_y. This draws attention to the nature of the variation $\mathcal{V}\gamma^1$ of any cycle γ^1 under the same conditions.

Draw lacets aa_k on S_y. Owing to (7, IV) as y describes aa_k a certain cycle δ_k^1 of H_y tends to the point of contact A_k of the plane $y = a_k$ and hence is ~ 0 on H_{a_k}. This is the *vanishing cycle* as $y \to a_k$. A simple lacet consideration shows that as y turns once positively around a_k the variation $\mathcal{V}\gamma^1$ of the cycle γ^1 is given by

$$\mathcal{V}\overset{1}{\gamma} = (\overset{1}{\gamma}, \overset{1}{\delta_k})\overset{1}{\delta_k}. \tag{8.2}$$

Hence

(8.3) THEOREM. *N.a.s.c. for invariance of the cycle* γ^1 *is that every*

$$(\overset{1}{\gamma}, \overset{1}{\delta_k}) = 0.$$

A noteworthy generalization is obtained when γ^1 is replaced by a one-chain L uniquely determined in term of y provided that y crosses no lacet.[2] As y turns as above around a_k the variation of L is

$$\mathcal{V}(L) = (L, \overset{1}{\delta_k})\overset{1}{\delta_k}. \tag{8.4}$$

Noteworthy special cases are

I. L is an oriented arc joining in H_y two fixed points of H_y.

II. Let C be an algebraic curve of F and let $M_1, \cdots, M_n$ be its intersections with H_y. Then L is a set of paths from a P_j to every point M_k in H_y.

(8.5) THEOREM. *The number of invariant cycles of* H_y *is equal to the Betti number* $R_1(F)$ *and both are even*: $r = R_1 = 2q$.

[2] In modern terminology, L will be a relative cycle.

This property was first proved in [7], although it was often admitted before. I give here an outline of the proof (not too different from the proof of [7]).[3]

To make the proof clearer I will use the following special notations: Γ a 3-cycle of F; $\{\Gamma_h\}$ base for the Γ's; $\gamma = \Gamma H_y$: (one-cycle of H_y); $\{\alpha_h\}$, $h \leqq 2p$, base for one-cycles of H_y; $\{\beta_j\}$, $j \leqq 2p-r$, base for the one cycles of H_y, none invariant; β any linear combination of the β_j;

Matrices such as $[(\beta_h, \Gamma_k)]_F$ will be written $[\beta\Gamma]_F$.

Proof that $r = R_1$. $\gamma = \Gamma H_y$ is invariant; conversely γ invariant is a ΓH_y. Moreover $\gamma \sim 0$ in H_y and $\Gamma \sim 0$ in F are equivalent. Hence $\{\gamma_h\}$, $h \leqq R_1$, is a base for invariant cycles and therefore $r = R_1$.

Proof that r is even. Since no β is invariant, $[\beta\delta]$ is of rank $2p-r$. Hence there exist $2p-r$ cycles δ which are independent in H_y. Denote them by $\bar{\delta}_h$, $h \leqq 2p-r$. Since $(\gamma_k \delta_h) = 0$ for every k, the δ_h depend on the β_j in H_y. Hence one may take $\{\gamma_h; \bar{\delta}_k\}$ as base for the one-cycles of H_y. Hence

$$[\gamma\delta] = \begin{bmatrix} [\gamma\gamma] & 0 \\ 0 & [\delta\delta] \end{bmatrix} \tag{8.6}$$

is nonsingular. It follows that $[\gamma\gamma]$ is likewise nonsingular. Since it is skew-symmetric, a well-known theorem of algebra states that r is even.

9. **The two-cycles of** F. From the expression (7.2) of the characteristic we have

$$\chi(F) = I + 4 = R_2 - 2R_1 + 2.$$

Hence

$$R_2 = I + 2R_1 + 2. \tag{9.1}$$

Besides this formula it is of interest to give an analysis of the 2-cycles.

Given a γ^2 one may assume it such that it meets every H_y in at most a finite set of points. Let Q be one of these and let P, Q be a directed path from the fixed point P to the point Q in H_y. Call L the sum of these paths. As y describes $S_y - \sum$ lacets aa_k, L generates a 3-chain C^3 whose boundary ∂C^3 consists of these chains:

(a) As y describes aa_k the vanishing one-cycle δ_k^1 of H_y generates a 2-chain Δ_k whose boundary

---

[3] The point here is to prove that an invariant cycle, which is also a vanishing cycle, is necessarily zero.

$$\partial\Delta_k = (\overset{1}{\delta}_k)_{H_a}.$$

The corresponding contribution to ∂C^3 is $\mu_k\Delta_k$, where (Zariski)

$$(9.2)\qquad \mu_1 = (L, \overset{1}{\delta}_1), \qquad \mu_k = (L + \mu_1\overset{1}{\delta}_2 + \cdots + \mu_{k-1}\overset{1}{\delta}_{k-1}, \overset{1}{\delta}_k).$$

(b) A part (H_a) of H_a.

(c) $-\gamma^2$ itself.

Hence

$$\partial C^3 = -\gamma^2 + \sum \mu_k\Delta_k + (H_a) \sim 0$$

and so

$$(9.3)\qquad \gamma^2 \sim \sum \mu_k\Delta_k + (H_a).$$

Since the right side is a cycle, and $y=a$ is arbitrary we have

$$(9.4)\qquad \sum \mu_k\overset{1}{\delta}_k \sim 0 \qquad \text{in } H_y.$$

Conversely when (9.4) holds, (9.3) is a 2-cycle. Thus to obtain R_2 it is merely necessary to compute the number of linearly independent relations (9.4) and add to them one unit for all μ_k zero, that is for the cycle H_a itself. This yields again (9.1).

For purposes of counting certain double integrals Picard required the number of *finite* 2-cycles independent relative to homologies in $F-H_\infty$. This is the number $R_2(F-H)$ and he found effectively

$$(9.5)\qquad R_2(F-H) = R_2 - 1.$$

10. **Topology of algebraic varieties.** I have dealt with it at length in both [8] and [9]. Questions of orientation and intersection are easily apprehended from the case of surfaces. I shall only recall here a few properties that are not immediate derivatives from the case of a surface.

The designations V^n, H_y are the same as in Chapter I. The following properties are taken from [9, Chapter V]. The symbol γ^k will represent a k-cycle of V^n.

I. *Every* γ^k, $k<n-1$, *of* H_y *is invariant.*

II. *Every* γ^k, $k<n$, *of* V^n *is* $\sim\gamma^{k'}$ *in* H_y.

III. *When* $k\leqq n-2$, $\gamma^k\sim 0$ *in* V^n *and* $\gamma^{k'}\sim 0$ *in* H_y *are equivalent relations.*

IV. *Under the same conditions* $R_k(V^n)=R_k(H_y)$.

III. Analysis with Little Topology

This is a rapid résumé of the extensive contributions of Picard, Severi and Poincaré upon which I applied topology (see IV). I will continue to consider the same surface F and all notations of II.

11. Émile Picard and differentials on a surface. During the period 1882–1906 Picard developed almost single-handedly the foundations of this theory. His evident purpose was to extend the Abel-Riemann theory and this he accomplished in large measure. Reference: Picard-Simart [**2**].

Picard studied particularly closed ω^1, that is

$$\omega^1 = \alpha dx + \beta dy, \qquad \partial\alpha/\partial y = \partial\beta/\partial x$$

and ω^2. The choice of closed ω^1 is very appropriate since then $\int\omega^1$ is an element of $K(F)$, and analytic function theory plus topology are fairly readily available.[4]

For closed one-forms the same three kinds as for abelian differentials are distinguished, save that for the third kind logarithmic curves replace logarithmic points.

Significant results are

I. Closed one-forms of the first kind make up a $\mathfrak{V}^q$ (Castelnuovo) ($q = \frac{1}{2}R_1$ as I have shown).

II. For the second kind same property save that they form a $\mathfrak{V}^{2q}$ mod $dK(F)$. (Picard)

III. Regarding the third kind Picard obtained this noteworthy result: There exists a least number $\rho \geqq 1$ such that any set of $\rho + 1$ curves are logarithmic for some closed ω^1 having no other poles.

The 2-forms admit again three kinds: (a) first kind: holomorphic everywhere; (b) second kind: holomorphic to within a $d\omega^1$ about each point; (c) the rest. The third kind is characterized by the possession of periods: *residues* over some 2-cycle γ^2 bounding an arbitrarily small neighborhood of a one-cycle on a curve.

The 2-forms of the first kind were already found by Max Noether. They are of the type

$$\omega^2 = \frac{Q(x, y, z)dxdy}{F_z'}$$

where Q is an adjoint polynomial of degree $m-4$. These ω^2 (or the associated Q) make up a $\mathfrak{V}^{p_g}$, where p_g is the *geometric* genus of F, studied at length by Italian geometers.

---

[4] Strictly speaking, $\int\omega^1$ is in $K(F)$ only if ω^1 has no residues or periods, but since $d\omega^1 = 0$, $\int\omega^1$ is invariant under a continuous variation in the path of integration.

Let $\mathfrak{V}^{\rho_0}$ be the vector space of the ω^2 of the second kind mod $d\omega^1$. Picard utilized his topological description of *finite* 2-cycles to arrive at the following formula:

$$\rho_0 = I + 4q - \rho + 2. \tag{11.1}$$

12. **Severi and the theory of the base.** The central idea here is a notion of *algebraic dependence* between curves on the surface F. I must first describe this concept.

Let the nonsingular surface F be in an S^{k+2}. A linear system of hypersurfaces of the space cuts out on F a *linear* system of curves $|C|$. This system is *complete* if its curves are not curves of an amplified linear system.

We owe to the Italian School the following property: Every sufficiently ample complete system $|C|$ is part of a collection $\{C\}$ of ∞^q such systems. The elements $|C|$ of the collection are in an algebraic one-one correspondence with the points of an abelian variety V^q, unique for F and called sometimes the *Picard variety* of F (see § 18).

A system $\{C\}$, ∞^2 at least, without fixed points and with irreducible generic curve is said to be *effective*. Its curves are also called *effective*.

Note the following properties:

(a) An effective system is fully individualized by any one of its curves.

(b) The generic curves of an effective system have the same genus, written $[C]$.

(c) The curves C, D of two effective systems intersect in a set of distinct points whose number is denoted by $[CD]$. In particular we write $[C^2]$ for $[CC]$ and $[C^2]$ is the *degree* of C.[5]

(d) With C, D as before let two curves C, D taken together be individuals of an effective system $\{A\}$. This system is unique and we write

$$A = B + C. \tag{12.1}$$

(e) Any two curves A_1, A_2 of an effective system $\{A\}$ may be joined in $\{A\}$ by a continuous system ∞^1 of curves of $\{A\}$, whose genus, except for those of A_1 and A_2, is fixed and equal to $[A]$.

As an application of (e) let A, B, C be effective and $A = B + C$. Following Enriques, connect A to $B + C$ as indicated in (e). There follows a relation

[5] This degree should not be confused with the degree of C as an algebraic curve in projective space.

$$\chi(A) + [BC] = \chi(B) + \chi(C) - [BC].$$

Hence if we define

$$\phi(A) = \chi(A) + [A^2] = 2 - 2[A] + [A^2],$$

we verify at once that

$$\phi(A) = \phi(B + C) = \phi(B) + \phi(C).$$

That is $\phi(A)$ is *an additive function on effective systems.*

When (12.1) holds between effective systems we set

$$C = A - B$$

and we have

$$\phi(C) = \phi(A - B) = \phi(A) - \phi(B).$$

Note also that as regards the symbols $[BC]$ we may operate as with numbers, that is

$$[(B \pm C)D] = [BD] \pm [CD].$$

Virtual systems: Let $\{A\}$, $\{B\}$ be effective systems. Without imposing any further condition define a virtual system $\{C\} = \{A - B\}$ as the pair of symbols $\{\phi(A) - \phi(B)\}$, $[(A-B)^2]$. This defines automatically $[C]$ and $[C^2]$. It is also clear that they are the same for $A-B$ and $A+D-(B+D)$ whatever D effective. In other words $\{C\}$ depends only upon the difference $A-B$. The symbol $\{C\}$ is called a virtual algebraic system of curves and $[C]$, $[C^2]$ are the related virtual genus and degree.

It may very well happen that while A, B are effective there exist curves C, not necessarily effective such that $B+C$ (B together with C) is a member of $\{A\}$. If so C is considered as a curve of the virtual system $\{C\}$ and has virtual characters $[C]$ and $[C^2]$, not necessarily its actual characters.

If we define $\{0\} = \{A - A\}$, as a virtual curve 0 is unique. One readily finds that $[0] = 1$, $[0^2] = 0$.

To sum up, the totality of effective and virtual curves form a module M_S over the integers: the Severi module. Within M_s a relation

$$\lambda_1 C_1 + \cdots + \lambda_s C_s = 0 \tag{12.2}$$

has a definite meaning. It is a relation of *algebraic dependence* between curves of F in the sense of Severi.

The following remarkable result was proved by Severi:

(12.3) THEOREM OF SEVERI. *The module of curves of F has a base consisting of ρ effective curves $C_1, C_2, \cdots, C_\rho$, where ρ is the Picard number relative to closed ω^1 of the third kind.*

That is any curve C satisfies a relation

$$\lambda C = \lambda_1 C_1 + \cdots + \lambda_\rho C_\rho$$

where λ and the λ_h are integers and $\lambda \neq 0$.

Severi also proved

(12.4) *The base may be chosen minimal, that is such that*

$$\lambda C = \lambda \sum \lambda_h C_h.$$

Moreover there exist effective curves $D_1, D_2, \cdots, D_{\sigma-1}$ such that actually

$$C = \sum \lambda_h C_h + \sum \mu_j D_j.$$

One assumes, as one may that σ is the least possible.

Severi also proved the following criteria:

(12.5) *A n.a.s.c. in order that the curves $C_1, C_2, \cdots, C_s$ be algebraically independent is that, with H a plane section, the matrix*

$$\begin{bmatrix} [C_h & C_k] \\ [C_h & H] \end{bmatrix}$$

be of rank s.

(12.6) *N.a.s.c. in order that $\{C_h\}$, $h \leqq \rho$, be a base is that the determinant $|[C_h C_k]| \neq 0$ and that its order ρ be the highest order for which this holds. Moreover, ρ is the Picard number.*

13. **Poincaré and normal functions.** Through an ingenious application of the theorems of Abel and Jacobi Poincaré arrived at a rapid derivation of some of the major results of Picard and the Italian geometers. I shall mainly deal with the part referring to Severi's theory of the base.

Let me first put in a most convenient form due to Picard and Castelnuovo the ω^1 of the first kind of the curve H_y. A base for them may be chosen of type

$$(13.1) \qquad du_s = \frac{Q_s(x, y, z)dx}{F'_z}, \qquad s \leqq p$$

where Q_s is an adjoint polynomial of degree $m-3$ in x and z. For the first $p-q$ the polynomial is of degree $m-3$ in x, y and z. For $s=p-q+1, \cdots, p$, it is of degree $m-2$ in x, y, z. Actually within this last range one may choose the Q_h so that the du_h only have constant periods relative to the invariant cycles and zero relative to the rest. As for the first $p-q$ they will have zero periods relative to the invariant cycles.

Let C be a curve on F and $M_1, \cdots, M_n$ its intersections with H_y. The sums from a fixed point P_1 of H_y to the M_k

$$\sum_k \int_{P_1}^{M_k} du_s = v_s, \qquad s \leqq p,$$

(integration in H_y) are Poincaré's normal functions.

Let L be the set of integration paths and with δ_k^1 as in (8) let

$$\Omega_{ks} = \int_{\delta_k^1} du_s. \tag{13.2}$$

Then with the μ_k as in (9.2) we find

$$\begin{aligned} v_h &= \sum \frac{\mu_k}{2\pi i}\int_a^{a_k} \frac{\Omega_{kh}(Y)dY}{Y-y} \qquad h = 1, 2, \cdots, p-q \\ v_{p-q+j} &= \alpha_j(\text{constant}) \qquad j = 1, 2, \cdots, q. \end{aligned} \tag{13.3}$$

REMARK. The only condition imposed upon the points M_k is that they be rationally defined together on H_y. They may represent for example the following special cases: (a) any sum of multiples of the fixed points P_h of H_y, in particular they may represent just μP_h; (b) if C is reducible say $C=C_1+C_2$ with M_{1h} and M_{2h} as respective intersections one might have any set $t_1\sum M_{1h}+t_2\sum M_{2h}$, and similarly for several reducible curves; (c) any combination of the preceding two cases. In what follows, "curve" must be understood to include all these special cases.

As usual when dealing with abelian sums the v_s are only determined mod periods of the related u_s.

(13.4) THEOREM OF POINCARÉ. *N.a.s.c. to have a set of v_s given by (13.3) represent a curve by means of Jacobi's inversion theorem are*

$$\sum_h \mu_h \Omega_{hs}(a) = 0, \qquad s = 1, 2, \cdots, p. \tag{13.4a}$$

(13.4b) *Let $P(x, y, z)$ be any linear combination of the $P_s(x, y, z)$ divisible by $y-a$ and let*

$$du = \frac{P(x, y, z)dx}{F_z'}, \qquad \Omega_k(y) = \int_{\delta_k^1} du.$$

Then one must have

$$\sum_{k} \mu_k \int_a^{a_k} \Omega_k(y)dy = 0. \tag{13.4c}$$

(13.5) *Comparison with Severi's results.* Let the collection $\{\mu_s\}$ of the μ_s occurring in any set of normal functions be designated by μ. The collection $\{\mu\}$ is a module U. Let U_0 be the submodule of all the elements μ^0 corresponding to the $\sum t_h P_h$, t_h an integer. The quotient $U_1 = U/U_0$ is the factor-module corresponding to all the curves which are not a plane section or more generally a $\sum t_j P_j$. The U_1 module has a base made up of $\rho - 1$ algebraically independent curves and a minimum base consisting of $\rho + \sigma - 2$ curves. By adding the $\mu(H)$ one has respectively ρ and $\rho + \sigma - 1$ for base and minimum base.

The quotient module $U_1 = U/U_0$ is the module of all μ of curves none a plane section H. The module $U_1 + H = M_p$ is the *Poincaré* module and it is isomorphic with the Severi module M_s.

(13.6) REMARK. In order to get rapidly to the "heart of the matter" I have assumed at the outset that in (13.1) the polynomials Q_{p-q+j} were of degree $m-2$ in x, y, z. This was based upon rather deep results of Picard and Castelnuovo. Poincaré however merely assumed that the degree of Q_{p-q+j} was $m-2+\nu_j$. As a consequence in (13.3) the constants α_j must be replaced by polynomials $\alpha_j(y)$ of degree ν_j. Then Poincaré shows that on the strength of the theorems of Abel and Jacobi every $\nu_j = 0$ hence the $\alpha_j(y)$ must be constants and one has in fact the form (13.3).

Notice also that from the form of the Q_{m-3+j} one may find another adjoint polynomial R_{m-3+j} of degree $m-3$ in y, z and $m-2$ in x, y, z such that

$$dw_j = \frac{Q_{m-3+j}dx + R_{m-3+j}dy}{F_z'}$$

is a closed ω^1 of the first kind. The set $\{dw_j\}$ is then shown to be a base for such differentials. This proves rapidly that their "independent number" is q. Finally since the α_j are arbitrary constants the form of (13.3) shows implicitly that a complete (maximal) algebraic system of curves consists of ∞^q linear systems in one-one correspondence with the points of an abelian q dimensional variety (see IV, §17).

In outline this shows how normal functions enabled Poincaré to obtain with ease a number of the major results of Picard and the Italian geometers.

IV. Analysis with Topology

14. **On the Betti number R_1.** In II I recalled my proof that R_1 is even and $R_1 = 2q$, the number of invariant cycles of the curve H_y. This gave incidentally a direct topological proof that the number of independent one-cycles in any curve of a sufficiently general system was fixed and equal to R_1. It showed also that the irregularity q of a surface, in the sense of Castelnuovo and Enriques was actually a topological character. As I will show in §17, a topological proof that q is an "absolute invariant" is immediate. Notice also that the distribution of complete algebraic systems in ∞^q linear systems, referred to in (13.6) is also shown to have topological character.

(14.1) Let $\{du_k\}$, $k \leqq q$, be a base for the closed ω^1 of the first kind of F. On H_y they coincide with the u_{p-q+h} of §13. Let $\pi_{k\mu}$, $\mu \leqq 2q$, be the periods of u_k relative to a homology base $\{\gamma_\mu^1\}$, $\mu \leqq 2q$, for the one-cycles of F. From the fact that the periods of the differentials of the first kind of H_y form a Riemann matrix, we infer:

(14.2) Theorem. *The matrix π of the periods of the u_k and their conjugates $\bar{u}_k$ as to the γ_μ^2 is a Riemann $2q \times 2q$ matrix.*

15. **On algebraic two-cycles.** A collection of mutually homologous 2-cycles is a *homology class.* In this manner algebraic cycles yield algebraic homology classes. Through addition they generate a module M_L. Thus in relation to the collection of curves on a surface F there are three definite modules: M_S (Severi module), M_P (Poincaré module) and M_L (Lefschetz module).

(15.1) Theorem. *The three modules M_S, M_P and M_L are identical.*

This property will be a final consequence of an extensive argument.

Returning to Poincaré's normal functions (III, §13) a glance at his two conditions for a set of normal functions to represent an algebraic curve reveals immediately that Poincaré's first condition simply means that

$$(15.2) \qquad \begin{cases} \gamma^2 = \sum \mu_k \Delta_k + (H_a) \\ \sum \mu_k \overset{1}{\delta}_k \sim 0 \qquad \text{in } H_y \end{cases}$$

is a cycle. As to the second condition it says merely that if

$$\omega^2 = \frac{Q(x, y, z)dxdy}{F_z'}$$

is of the first kind, that is if Q is adjoint of order $m-4$, then

$$\int_{\gamma 2} \omega^2 = 0.$$

Hence Poincaré's conditions are equivalent to the following result:

(15.3) THEOREM. *A n.a.s.c. for a cycle γ^2 to be algebraic is that the period of every 2-form of the first kind relative to γ^2 be zero.*

(15.4) REMARK. Among all the "algebraic" curves there were included all the sums $\sum m_j P_j$, where the P_j are the fixed points of H_y. It is evident that for these special "2-cycles" $\int \omega^2$ is zero.

(15.5) COROLLARY. *Severi's number σ is merely the order of the torsion group of the two-cycles (or equally of the torsion group of the one-cycles).*

For if γ^2 is a torsion 2-cycle we have $\lambda\gamma^2 \sim 0$, $\lambda \neq 0$, and hence

$$\int_{\gamma 2} \omega^2 = 0$$

for every ω^2 of the first kind.

(15.6) THEOREM. *The number ρ is the Betti number of algebraic cycles.*

This is a consequence of the following property:

(15.7) *Let $C_1, \cdots, C_s$ be a set of curves and let $\overline{C}_h$ be the cycle of C_h. Then*

P_a: *algebraic independence of the C_h*

P_h: *homology independence of the $\overline{C}_h$*

are equivalent properties.

From obvious considerations P_a implies P_h. Conversely let P_h hold. We must show that $\overline{C} \sim 0$ implies $\lambda C = 0$, $\lambda \neq 0$. Here I follow Albanese's rapid argument. Let $C = A - B$, A and B effective. Since $\overline{A} \sim \overline{B}$ and $[CD] = (\overline{C}, \overline{D})$ we have

$$[A^2] = [AB] = [B^2]; \qquad [AH] = [BH]$$

where H is a plane section. Hence Severi's independence criterion is violated between A and B. Consequently $\lambda A = \mu B$, $\lambda\mu \neq 0$. From $[AH] = [BH]$ follows $\lambda = \mu$ and therefore $\lambda(A - B) = 0 = \lambda C$, $\lambda \neq 0$. This proves (15.7).

It follows that $M_S = M_L$ and as $M_P = M_S$, (15.1) is proved.

Notice that we may now give the following *very simple* definition of virtual curve: it is merely an algebraic 2-cycle. Simplicity is even augmentable by replacing everywhere the symbol $=$ of algebraic dependence ($=$) by the homology symbol $\sim$.

16. **On 2-forms of the second kind.** The basic result is the proof of the formula

$$\rho_0 = R_2 - \rho. \tag{16.1}$$

I shall just indicate an outline of my proof. I shall also show that the process outlined obtains incidentally Picard's fundamental result for ρ concerning logarithmic curves of a closed ω^1 of third kind. The steps follow closely an analogous outline in my monograph [9].

For convenience I call ω^1 and ω^2 *regular* when

$$\omega^1 = \frac{Pdy - Qdx}{\phi(y)F_z'}, \qquad \omega^2 = \frac{Pdxdy}{\phi(y)F_z'},$$

where P and Q are adjoint polynomials and $\phi(y)$ is a polynomial.

If $\omega^2 = d\omega^1$, ω_2 is said to be *improper*. Thus ρ_0 is the dimension of the vector space of the ω^2 of the second kind mod those which are improper.

By *reduction* of ω^2 I understand the subtraction of an improper ω^2.

I. *The periods and residues of a normal 2-form are arbitrary.*

II. *One may reduce any ω^2 of the second kind to the regular type.*

III. *A regular ω^2 such that $\int\omega^2$ has neither residues nor periods is reducible to a regular $d\omega^1$.*

Except for the presence of the polynomial $\phi(y)$ the proofs of the preceding propositions are very close to those of Picard. It is true that allowing $\phi(y)$ in regular ω^1 and ω^2 considerably simplifies every step (see [9, Note I]).

IV. *Let C be a curve of order s. We may choose coordinates such that C does not pass through any of the fixed points P_j of H_y, nor through the points of contact of the planes $y = a_k$. One may form an $\omega^1 = Rdx$, $R \in K(F)$ possessing on H_y the s-logarithmic points of CH_y with logarithmic period $2\pi i$ and say P_1 with logarithmic period $-2\pi is$. One may even select R so that $(\partial R/\partial y)dx$ has no periods. From this follows that there is an $S(x, y, z) \in K(F)$ such that*

$$\omega^2 = d(Rdx + Sdy)$$

is regular.

Take now $C_1, C_2, \cdots, C_t$ and the axes so chosen that they all behave like C. Let ω_h^2 be analagous to ω^2 for C_h.

Owing to III *it is now readily shown that n.a.s.c. in order that some linear combination*

$$\omega^2 = \sum \alpha_h \omega_h^1$$

be without periods is that the C_s *and* H *be logarithmic curves of a closed* ω^1.

Since R_2 is finite there is a least $\rho-1$ such that for $s=\rho$ the curve C_h, H are logarithmic curves of a closed ω^1 of the third kind. Hence

(16.2) *Picard's fundamental property for* ρ *is a consequence of the finiteness of the Betti number* R_2.

V. *To proceed one may form* $\rho-1$ *linearly independent* ω^2 *which are improper. Since the total number of distinct periods is equal to* $R_2(F-H)$ $=R_2-1$ *we have then* $\rho_0=R_2-\rho$, *as asserted.*

(16.3) *On Picard's treatment of* ρ_0 *and* ρ. Owing to lack of topological technique Picard proved directly that ρ_0 was finite by showing through strong algebraic operations that if

$$\omega^2 = \frac{Q(x, y, z)dxdy}{F_z'}$$

where Q is adjoint, is of the second kind, the degree of Q was bounded.

Although Picard did not observe it, his later treatment of ω^2 of the second kind contained implicitly (argument of 16.2) the proof that ρ had the property by which he defined it relative to closed ω^1 of the third kind.

17. Absolute and relative birational invariance. Take again a general n-variety

$$F(x_1, \cdots, x_n, y) = 0. \tag{17.1}$$

Let $\{\xi_0, \cdots, \xi_r\}$ be a homogeneous base for the function field $K(F)$. Then the system

$$\tau y_h = \xi_h, \qquad h = 0, 1, 2, \cdots, r > n$$

represents a model F_1 of F in the projective space S^r, with homogeneous coordinates y_h. If $\{\eta_0, \cdots, \eta_s\}$, $s>n$, is a second homogeneous base for $K(F)$, the system

$$\sigma z_k = \eta_k, \qquad k = 0, 1, \cdots, s$$

represents a second model F_2 of F in a projective space S^s. Since $\{\xi_h\}$ and $\{\eta_k\}$ are homogeneous bases for $K(F)$ F_1 and F_2 are birationally transformable into one another. The simple example of two elliptic curves of degrees 3 and 4 show however that the corresponding structures need not be homeomorphic. The difficulty is caused by the presence of singularities. A standard device for curves enables one to "forget" singularities and restore homeomorphism. No such device is known for a V^n, $n>1$.

For simplicity let me limit the argument to surfaces. I have really considered a surface as a nonsingular model in some projective space. Let F_1, F_2 be two such distinct models and suppose that the *field $K(F)$ is not that of a ruled surface.* Then according to Castelnuovo and Enriques a birational transformation T: $F_1 \rightarrow F_2$ may take a finite number δ_{12} of *exceptional* points of F into *disjoint* nonsingular rational curves. There exists an analogous δ_{21} for T^{-1}. Let a point P of F_1 be sent by T into a curve C of F_2. Since C is rational and nonsingular it is topologically a sphere. Hence its characteristic $\chi(C)=2$. Hence the gain in $\chi(F_1)$ through δ_{12} exceptional points is δ_{12}. Therefore

$$\chi(F_1) + \delta_{12} = \chi(F_2) + \delta_{21}. \tag{17.2}$$

Now a character, numerical or other of F is said to be an *absolute* invariant if it is unchanged under all transformations such as T. A *relative invariant* is one that may change under certain transformations T.

Let me examine some of the characters that have been introduced.

It is readily shown that under T both R_2 and ρ are increased by the same amount $\delta_{12}-\delta_{21}$. Hence both are relative invariants and $\rho_0=R_2-\rho$ is an absolute invariant.

Since

$$\chi(F) = R_2 - 2R_1 + 2$$

and both χ and R_2 vary in the same way, χ is a relative invariant and R_1 is an absolute invariant.

Therefore:

(17.3) *The dimensions of the spaces of closed ω^1 of the first and second kinds and of ω^2 of the second kind are absolute invariants.*

18. **Application to abelian varieties.** Let Π and M be a Riemann matrix and its principal matrix (see §3).

Introduce the following vectors:

$$\boldsymbol{u} = (u_1, \cdots, u_{2p}), \qquad u_{p+j} = \bar{u}_j$$

$$\pi_\mu = (\pi_{1\mu}, \cdots, \pi_{2p,\mu}), \qquad \mu = 1, 2, \cdots, 2p; \qquad \pi_{p+j,\mu} = \bar{\pi}_{j\mu}.$$

Through the hyperplanes

$$u = \sum s_\mu \pi_\mu,$$

s_μ an integer, real $2p$-space is partitioned in a familiar way into paralellotopes. A suitable fundamental domain D is

$$u = \sum t_\mu \pi_\mu, \qquad 0 \leqq t_\mu < 1.$$

The identification of congruent boundary points turns this domain into a $2p$-ring R^{2p} (product of $2p$ circles):

Corresponding to Π and M there may be defined a whole family of functions θ of various orders. Each such function ϕ is a holomorphic function in the domain D. Those of a given order, say n, are characterized by the property that $\phi_n(u+\pi_\mu)=\phi_n(u)$ times a fixed linear exponential function of u. I have shown that one may find an n such that if $\{\theta_n^j(u+\alpha)\}$, α a fixed p-vector, $j=0, 1, \cdots, r$, is a finite linear base for the $\theta_n(u+\alpha)$ then the system

$$kx_j = \overset{j}{\theta}_n(u+\alpha)$$

represents a nonsingular p-variety V^p in projective S^r, and this V^p is in analytic homeomorphism with the ring R^{2p}. This is an abelian p-variety (see [8])

The topological relation $V^p \leftrightarrow R^{2p}$ assigns an exceptionally simple topology to V^p. Let the edges of D oriented from the origin out be designated by $1, 2, \cdots, 2p$. Any i_h defines a one-cycle represented by (i_h); any two edges i_h, i_k define a 2-cycle represented by (i_h, i_k), etc. The (i_h, i_k), $i_h < i_k$ form a base for the 2-cycles of V^p, etc.

I am mainly concerned with the 2-cycles. In view of $(\nu, \mu) = -(\mu, \nu)$ a general 2-cycle is represented by a homology

$$\gamma^2 \sim \sum m_{\mu\nu}(\mu, \nu), \qquad m_{\mu\nu} = -m_{\nu\mu}.$$

On the other hand

$$\overset{2}{\omega}_{jk} = du_j du_k; \qquad j, k \leqq p; \qquad j < k$$

is a closed 2-form of the first kind of V^p and $\{du_h du_k\}$ is a base for all such forms.

(18.2) REMARKS. *On a general n-variety* V^n, $n \geqq 2$. Considerations of the same type as in §12 may be extended automatically to algebraic dependence of hypersurfaces of V^n (its V^{n-1}), and also to their $(2n-2)$-cycles. Algebraic and homology dependence give rise to a number $\rho(V^n)$. I single out especially the following proposition from [9, p. 104] (Corollary):

(18.3) THEOREM. *Let* Φ *be a fixed surface of* V^n *(general intersection of hyperplane sections of* V^n*) and let* $C_1, \cdots, C_s$ *be hypersurfaces which cut* Φ *in curves* C_h^*, $h \leqq s$. *Then the following relations are all equivalent: relations of algebraic dependence between the* C_h *in* V^n, *the same between the* C_h^* *in* Φ; *relations of homology between the standard oriented cycles* $\overline{C}_h$, C_h *in* V^n; *the same for the* C_h^* *in* Φ.

Returning now to the abelian variety V^p let Φ, C, C^* be this time the same as above but for V^p. Now the ω_{hk}^2 taken on Φ become ω^2 of the first kind for Φ. If $\{C_s\}$, $s \leqq \rho(V^p)$, is a base as to $=$, or equivalently as to $\sim$ and algebraic $(2p-2)$-cycles of V^p, then the same holds for the curves C_s^* in Φ. Hence by theorem (15.2):

$$\int_{C_s^*} du_j du_l = 0; \qquad j, l \leqq p. \tag{18.4}$$

On the other hand since the (μ, ν) are cycles in Φ we have in Φ

$$C_s^* \sim \sum m_{\mu\nu}^s (\mu, \nu), \qquad m_{\mu\nu} = -m_{\nu\mu}.$$

Hence (18.4) yields

$$\sum m_{\mu\nu}^s \pi_{j\mu} \pi_{k\nu} = 0; \qquad j, k \leqq p. \tag{18.5}$$

This really means that the ρ matrices $[m_{\mu\nu}^s]$ are linearly independent singularity matrices for the Riemann matrix Π. If the singularity index of Π is k, then one must have

$$\rho \leqq k. \tag{18.6}$$

The possible inequality is due to the fact that an algebraic 2-cycle of Φ must satisfy a relation such as (18.4) not merely with respect to the closed ω^2 of the first kind of V^p in Φ, but also with respect to all ω^2 of the first kind of Φ, and one cannot exclude the possible existence of such ω^2 other than the closed taken on Φ. However, the following two properties hold:

(a) There is a base for the forms M made up of principal forms.

(b) Each principal M gives rise to a particular system of functions ϕ à la θ. These functions are said to be *intermediary*.

(c) If $\{M_j\}$, $j \leqq k$, is a base for the matrices M, and ϕ_j is an intermediary function relative to M_j then $\phi_j = 0$ represents a hypersurface of V^p and these hypersurfaces are algebraically independent.

It follows that $\rho \geqq k$ and therefore

$$\rho = k. \tag{18.7}$$

This is the result that I was looking for.

Actually the relations between the hypersurfaces *as* cycles and their Severi independence are the same as for their sections with the surface Φ. That is,

(18.8) THEOREM. *For hypersurfaces of V^p algebraic dependence and homology in V^p are equivalent relations.*

BIBLIOGRAPHY

1. Émile Picard, *Triaté d'analyse*. Vol. 2, Gauthier-Villars, Paris.
2. Émile Picard and George Simart, *Théorie des fonctions algébriques de deux variables indépendantes*. Vols. I, II, Gauthier-Villars, Paris, 1895, 1906.
3. Francisco Severi, *Sulla totalità delle curve algebriche traceiate sopra una superficie algebrica*, Math. Ann. **62** (1906), 194–226.
4. ———, *La base minima pour la totalité des courbes tracées sur une surface algébrique*, Ann. Sci. École Norm. Sup. **25** (1908), 449–468.
5. Henri Poincaré, *Sur les courbes tracées sur une surface algébrique*, Ann. Sci. École Norm. Sup. **27** (1910), 55–108.
6. Gaetano Scorza, *Intorno alla teoria generale delle matrici di Riemann e ad alcune sus applicaciones*, Rend. Circ. Mat. Palermo **41** (1916), 263–380.
7. Solomon Lefschetz, *Algebraic surfaces, their cycles and integrals*, Ann. of Math. **21** (1920), 225–258.
8. ———, *On certain numerical invariants of algebraic varieties with application to abelian varieties*, Trans. Amer. Math. Soc. **22** (1921), 327–482.
9. ———, *L'analysis situs et la géométrie algébrique*, Borel Series, 1924.
10. ———, *Topology*, Amer. Math. Soc. Colloq. Publ., Vol. 12, Amer. Math. Soc., Providence, R. I., 1930; reprint Chelsea, New York, 2nd ed., 1950.
11. Oscar Zariski, *Algebraic surfaces*, Ergebnisse der Math., Springer-Verlag, Berlin, 1935; reprint Chelsea, New York, 1948.
12. Heisuki Hironaka, *Resolution of singularities of an algebraic variety over a field of characteristic zero*, Ann. of Math. (2) **79** (1964), 109–329.
13. W. V. D. Hodge, *Theory and application of harmonic integrals*, Cambridge Univ. Press, New York, 1941.

Contemporary Mathematics
Volume 58, Part I, 1986

SOLOMON LEFSCHETZ

SIR WILLIAM HODGE, F.R.S.†

Solomon Lefschetz, who died on 6 October, 1972 at the age of 88, was a dominant figure in the mathematical world, not only for his outstanding original contributions to at least three branches of mathematics, but also for his personal influence in creating world famous centres of mathematics at Princeton, Mexico and elsewhere. All this was achieved in spite of a crippling handicap caused by an accident in 1910 which forced him to abandon his chosen career as an engineer and begin again from scratch as a pure mathematician.

Personal

Solomon Lefschetz was born in Moscow on 3 September, 1884. His father's business required him to spend much time in Persia, and he decided to settle his family in Paris, where his children were brought up from a very early age. Young Sol's first language was French, and he only learned Russian as a teenager.

There is little on record about his early years in Paris, and the first event that is known was the award of the degree of " Ingénieur des arts et manufactures " in 1905, after he had spent 3 years at the École Centrale in Paris. Shortly after this he emigrated to the United States, and spent a few months at the Baldwin Locomotive works, before becoming an engineering apprentice, and later a member of the engineering staff, of the Westinghouse Electric and Manufacturing Co. of Pittsburg. He was with this firm from 1907 to 1910, but then what might well have become a highly successful career in industry was abruptly terminated by an accident at work in which Lefschetz lost both his hands.

After a period in hospital, he faced up to the fact that his career as an engineer was finished. He decided to make his career in pure mathematics, and with this in view he became a graduate student at Clark University (Worcester, Mass.) where he took his Ph.D. degree in 1911. He then occupied a series of positions in mid-western universities: instructor at the University of Nebraska (1911–13); instructor (1913–16), assistant professor (1916–19), associate professor (1919–23), and professor (1923–25), at the University of Kansas. It was during these 14 years that he came to terms with his disability, and re-gaining his self-confidence laid the foundations of his new career

† This Notice is reproduced from that prepared by Sir William Hodge for the Royal Society.

[Bull. London Math. Soc., 6 (1974), 198–217]

He became an American citizen in June 1912, and in the following year he married Miss Alice Berg Hayes, whom he met while at Clark University, and who was to be a tower of strength to him for the rest of his life. Mrs Oswald Veblen once told me that she first met Lefschetz shortly after his accident: he then seemed a poor downcast creature, with only rudimentary artificial hands. Mrs. Lefschetz helped him to overcome his despair and face up to life. Even after he had got over the initial difficulties she was an invaluable aid. When he had mastered the art of living with his disability, and had more efficient artificial limbs, he could still have accidents, and to see her comfort him and rally him was something never to be forgotten. Later still, his exuberance sometimes burst all bounds, and she was equally successful in quietening him down. She was, indeed, so successful with him that in later years his boundless energy was almost too much for her, and she found it too exhausting to accompany him on his many dashes by airplane about the world.

The years spent in Nebraska and Kansas were one of the most important periods of Lefschetz's life; not only did he learn to overcome his physical disability then, but the main part of his massive contributions to algebraic geometry was completed before he left Kansas for Princeton. It is not too much to say that he arrived in the middle-west an unknown person, and left 14 years later as one of the most outstanding geometers of the day. As he has said in *A page of mathematical autobiography*, his mathematical isolation was complete, and this circumstance was most valuable in that it enabled him to develop his ideas in complete mathematical calm.

His earliest publications dealt mainly with properties of algebraic loci in projective spaces, but quite early (by 1915) he demonstrated his real interest in the intrinsic properties of algebraic loci. His attention was first turned in this direction by reading E. Picard's *Traité d'analyse*, and *Théorie des fonctions algébriques de deux variables indépendantes* by E. Picard and G. Simart. In his *Page of mathematical autobiography*, Lefschetz explains how he came to apply topological methods to the theory of algebraic surfaces. Using Picard's number ρ_0 which he had come to believe was equal to the second Betti number of a surface, he computed this Betti number for a hyperelliptic surface, obtaining the value 5; whereas a simple direct calculation gave the Betti number as 6. He puzzled over this discrepancy for some time before he realized that Picard only dealt with finite 2-cycles whereas the section at infinity was also a cycle of the surface. This led him to consider the cycles carried by algebraic curves on a surface, and he realized that a full topological study of the topology of a surface would yield a rich field of geometrical results. As he put it: " The harpoon of algebraic topology was planted in the body of the whale of algebraic geometry."

In addition to the work of Picard referred to above, Lefschetz used a paper by H. Poincaré, " Sur les courbes tracées sur une surface algébrique " (*Ann. del'École Norm. Sup.* **27**, 1910) to great effect, and he was able to produce an essentially complete topological theory of algebraic surfaces, a fundamental aspect of which was the relation of his theory to F. Severi's theory of the base. This work was published in a lengthy series of papers, but the definitive account was published in the *Transactions of the American Mathematical Society* in 1921 [**24**]. For this he was awarded the Bôcher Memorial Prize of the American Mathematical Society. This was, in fact, a translation, slightly modified, of a memoir for which he received the Prix Bordin for 1919 from the Académie des Sciences, Paris. In 1924 he published a tract: *L'analyse situs et le géométrie algébrique* **[29]** in the Borel Series which served to make his work known to a much wider audience.

By this time Lefschetz was recognized as a powerful figure in the mathematical world, and he was receiving many invitations to visit other universities. In 1924 he was invited to spend a year as visiting professor at Princeton, at the end of which he was appointed to a permanent post there as associate professor. Three years later, in 1928, he was made a full professor, and in 1932 he succeeded Oswald Veblen, who had gone to the Institute of Advanced Study, as Henry B. Fine Research Professor, an office he held until his retirement in 1953.

The move to Princeton was a landmark in Lefschetz's life. The isolation of Kansas was over and he found himself in close contact with a wide circle of mathematicians on the permanent staff at Princeton, among the many distinguished visitors who spent periods of leave there, and among the graduate students, many of whom were to go on to great achievements. It was also easier for him to travel and visit other universities. That he was able to take full advantage of these opportunities in spite of his physical handicap was due to his indomitable courage. His friends were, of course, watchful in trying to avoid a catastrophe, but the number of occasions when they had to take any positive action was very small. In fact, the only instance I am aware of in which the University made special arrangements for him was the provision of a special lock on his office so that he could lock or unlock it without having to use a key; the value of this provision was however somewhat reduced because he was so pleased with the device that he kept showing all and sundry how to operate the secret button. The way in which he could look after himself, and lead a normal life, never ceased to astound all who knew him. He could travel by himself, and live in hotels or with friends without calling for assistance, and although his sporting activities were necessarily limited, he was quite a powerful swimmer. Socially, he was an exuberant but entertaining companion, and the only occasions for embarrassment that sometimes arose were when one introduced him to a stranger who was not expecting artificial hands.

When Lefschetz moved to Princeton he was nearly at the end of his main creative work on algebraic geometry. His main contribution to geometry after this was his paper [37] on correspondences between algebraic curves, but he kept in close touch with others who were working in the same field, and he frequently lectured on algebraic geometry and published improved and modernized accounts of his earlier work. But his activity now became centred on algebraic topology. Amongst the numerous distinguished mathematicians who were around Princeton when he arrived, the two with whom he was most closely associated were Oswald Veblen and J. W. Alexander. The interests of the latter were very close to those of Lefschetz, and although they never wrote a joint paper they frequently discussed such matters of common interest as fixed-point theory and duality theorems. Lefschetz was a great admirer of Alexander, and in later years was saddened when Alexander gradually withdrew from contact with mathematicians and became a recluse.

Lefschetz's original contributions to topology continued until about 1940, and will be described later. But possibly his main contribution to mathematics during the thirties lay in his powerful influence on others: he worked very hard to keep himself informed on what his students and associates were doing, and was a vigorous critic of anything he did not approve of. He asserted (with much truth) that "he made up his mind in a flash, and then found his reasons". Naturally he made mistakes this way, but once he was really convinced that he was wrong he could be extremely generous. This can be illustrated from the occasion when I first met him. His paper [37] on correspondences between curves contained *inter alia* a very simple proof of the Riemann inequalities for the integrals of the first kind on a curve. About the time it was written Severi and other Italian geometers were concerned with the question whether a double integral of the first kind could have all its periods zero, and I noticed that Lefschetz's method of dealing with the Riemann inequalities could answer this question at once; and I wrote a short paper on it. When this appeared Lefschetz "made up his mind in a flash"—I was wrong and I must withdraw the paper. A correspondence lasting for several months took place, and during this period Lefschetz was travelling round Europe visiting mathematicians in various countries to whom he voiced his criticisms of my paper. Eventually we reached a state of armed neutrality, and he extended an invitation to me to visit Princeton. When I arrived there I was immediately instructed to conduct a seminar on my paper (owing to his characteristic interruptions the seminar actually lasted for six sessions), at the end of which he stood up and publicly retracted all his criticisms, and then, despite his handicap, wrote to all his European correspondents admitting that I was right and he was wrong.

This method of organizing seminars on topics in which he was interested and then continually heckling the speaker was typical of the way in which he got things done. It was somewhat harassing for a young man not accustomed to it, but it was kindly meant, and often helpful. Eventually it became so famous that it earned Lefschetz a verse in the song sung by Princeton students about members of the Faculty:

Here's to Lefschetz (Solomon L.)
Who's as argumentative as hell,
When he's at last beneath the sod,
Then he'll start to heckle God.

He employed equally drastic methods in his capacity as editor of the *Annals of Mathematics* over a period of 25 years. No leniency was shown towards any paper submitted to the journal which was not up to his standards, and anyone who disagreed with his judgement had to work very hard to make him change his mind. But once he had decided that man was worth helping there was no limit to the aid he would give him. By these methods he made the *Annals* one of the top mathematical journals in the world, and he and his colleagues made Princeton a world centre of mathematics. In the course of this vigorous programme he made very few enemies indeed: one felt that there was no personal animosity in his bark, and no self-seeking: he just wanted to serve mathematics as best he could. His contribution to mathematics was recognized by his election to the National Academy of Sciences in 1925, and to the Presidency of the American Mathematical Society for 1935–36.

He became Chairman of the Department of Mathematics at Princeton in 1945. This marked another major change in his activities. In 1945–46, and again in 1947 he was an exchange professor at the National University of Mexico, and subsequently he paid many visits, particularly after his retirement from Princeton, to Mexico City. His enthusiasm, drive and organizing ability contributed greatly to the establishment of a lively school of mathematics there. In recognition of his services the Mexican government bestowed on him the Order of the Aztec Eagle.

During World War II Lefschetz served as a consultant to the U.S. Department of the Navy at the David Taylor Model Basin, where he came across the Russian research and writings on nonlinear oscillations and stability. He realized immediately the importance of the work of Poincaré and Liapunov on the geometrical theory of differential equations, and saw that the subject had been " too long neglected " in the United States. With the support of the Office of Naval Research (and against the advice of some of his colleagues who felt that federal support would endanger the freedom of science in the country) he organized (in 1946) a differential equations project at Princeton University, which became the leading centre of research in ordinary differential equations in the United States.

Lefschetz remained the director of this project until his retirement in 1953. During the next five years the Princeton project was gradually phased out, and he tried more than once unsuccessfully to have a research institution established in another American university. He became a consultant to R.I.A.S. (Research Institute for Advanced Studies) which was formed by the Martin Company in Baltimore as an experiment in the support of basic research by industry. In November 1957 the President and Board of Directors of the Martin company gave him *carte blanche* to establish at R.I.A.S. a centre for differential equations charging him to make "this centre an outstanding example of its kind in the world ". In three years the centre had gained an international reputation for its research on differential equations and on the mathematical theory of optimal control and stability. In 1964 the major portion of the R.I.A.S. Centre for differential equations moved to Brown University in Providence, Rhode Island, where they established within the Division of Applied Mathematics the Lefschetz Centre for Dynamical Systems. Brown University made him a visiting Professor of Applied Mathematics, and for six years he commuted weekly by air from Princeton to Providence, where he lectured, discussed research, and spread his wit, enthusiasm, and love of life and mathematics. A colleague who was closely associated with him aptly sums him up in the sentences: " Lefschetz was not only a great mathematician: he was a research administrator of the highest order. He was direct, decisive, and persuasive. He saw it as his duty to provide the atmosphere, environment, and conditions conducive to the best research. He could encourage and cajole, and when necessary he could rebuke. He never procrastinated, and his decisions and judgements were often instantaneous. He recognized the organizational importance of good secretaries, and they loved him and were devoted to him." All this is equally true of him as a professor at Princeton.

Lefschetz died after a short illness in Princeton on 6 October, 1972, leaving a widow in Princeton and a brother in Paris.

Many honours came to him. In addition to those already mentioned, he received the Antonio Feltrinelli International Prize of the Accademia Nazionale dei Lincei in 1956. In 1954, his colleagues in Princeton, in order to mark his seventieth birthday, organized an international symposium on algebraic geometry and topology, and in 1965 an international symposium on differential equations and dynamical systems was held at the University of Puerto Rico as a tribute to him. He received honorary degrees from the universities of Prague, Paris and Princeton, and Brown and Clark. In addition to the National Academy and the American Mathematical Society he was a member of the American Philosophical Society. He was an honorary member of many mathematical societies in Europe and America, and a foreign member of L'Académie des Sciences of Paris, the Academia Real de Ciencias in Madrid, the Reale Istituto Lombardo of Milan, of the Royal Society and of the London Mathematical Society. And in 1964 he received from President Lyndon B. Johnson the National Medal of Science "for indomitable leadership in developing mathematics and training mathematicians, for fundamental publications in algebraic geometry and topology, and for stimulating needed research in nonlinear control processes".

Scientific work

Lefschetz's contributions to mathematics were spread over 60 years, and his publications in one form or another number some 134 items. Most of his work was carried out in centres of almost feverish activity, and in order to give a complete account of his contributions to mathematics it would be necessary to write a history of pure mathematics over half a century. This is not possible within the scope of the present memoir, and we must confine ourselves to a general account of his work in three fields: algebraic geometry, algebraic topology, and differential equations. It is to be observed that his efforts in these fields were largely concentrated in different stages of his life, and overlapped very little. Thus his major contributions to algebraic geometry belong to the time he spent in Nebraska and Kansas (1911–25), his work in algebraic topology comes from the period stretching from his arrival in Princeton to the entry of the United States into World War II, and from the end of the war until his death his main concern was with differential equations. These dates are, of course, only approximate; his major contributions to algebraic geometry continued until about 1929, whereas his first contribution to algebraic topology is dated 1923. The overlap of his interests in topology and differential equations is not so marked. But as he moved on from one field to another, he never forgot his earlier loves, and in his later periods he wrote a substantial number of books, review articles, and memoirs on algebraic geometry and topology in which he gave fresh interpretations of his early work and did much to bring it up to date.

This tendency to channel his energies in one direction at a time can be explained as follows. He would discover in the writings of a predecessor an idea which was capable of an expansion undreamt of by the originator, and, developing this to its full potential, he would use it to make a major breakthrough in geometry or topology.

The exploitation of the idea would occupy him over a long period, and after he had achieved the central new contribution he would consider its application to particular problems. But quite soon he would become impatient of minor applications, and a great new idea would catch his attention, and his ability to concentrate all his efforts on a single idea would drive everything else out of his mind. This did not, however, prevent him from taking an interest in what others were doing with his earlier discoveries; but while he would follow this, and refer to it in survey articles, he never returned to the front line of the battle.

(*a*) *Algebraic geometry*

Lefschetz's contributions to algebraic geometry began in 1912. His first papers dealt with a variety of problems covering loci in projective space. But it was not until 1915 that he made contact with the field which he was destined to revolutionize. From 1915 onwards there appeared a succession of short papers recording the development of his ideas. The main results recorded in the papers appeared, amended and improved, in 1921 [**24**] and 1924 [**29**] (his Borel Tract), and these memoirs can be taken as the text of an account of this phase of his work.

The early development of algebraic geometry, as we know it, goes back to Abel, Riemann, and Weierstrass, who built up a theory of functions associated with an irreducible polynomial equation

$$f(x, y) = 0.$$

This was simply a branch of the theory of functions, and the outstanding feature of it was the introduction of the topological concept of the Riemann surface. About 1870, Max Noether began developing an alternative theory by considering $f(x, y) = 0$ as the equation of a curve in a plane, and using geometrical methods to obtain the same results (see Brill–Noether, *Math. Ann. J.* 1874, 269–310). In 1892 and 1893 G. Castelnuovo and F. Enriques began using the geometrical methods of Noether to construct a theory of algebraic surfaces: this was the foundation of the great " Italian " school of algebraic geometry. At the same time various French mathematicians (and also Noether himself) began studying the transcendental theory of surfaces. The results of their researches are contained in the treatise of E. Picard & G. Simart *Théorie des fonctions algébriques de deux variables indépendantes* (Paris, 1895, 1906). Unfortunately, the authors were severely handicapped by the absence of any powerful topological tools they could use, but in the face of this handicap their achievement was really remarkable. It was here that Lefschetz was able to make his great breakthrough; starting with the rather primitive topological work of Henri Poincaré, he was able to develop this and apply it to Picard's work, and so pave the way to a further burst of activity.

The origins of Lefschetz's ideas on the transcendental theory of algebraic varieties are scattered over a considerable number of papers, and as his investigations progressed he combined these with the other results due to Picard and others into an intricate pattern in such a way that it is not possible to give here a strictly chronological account of the work. References are therefore only given to publications of special significance. The authoritative account of his general theory of varieties is in [29], which contains numerous references to earlier work.

The transcendental theory of surfaces with which Picard and Simart were concerned was a natural generalization of the theory of algebraic integrals on the Riemann surface of a curve. This requires as a preliminary step the construction of a four-dimensional manifold $M(F)$ related to a surface F defined over the complex field as a Riemann surface is related to a curve. Certain preliminary difficulties arising from the singularities of F have to be overcome, but it is possible to construct a manifold $M(F)$ which represents F in the obvious way, which can be covered by a finite number of open sets U_α, each of which can be mapped by complex parameters onto a cell of the complex space (x_α, y_α). The transformation equations of the local coordinates $x_\alpha = f_{\alpha\beta}(x_\beta, y_\beta)$, $y_\beta = y_{\alpha\beta}(x_\beta, y_\beta)$ are analytic and non-singular. The functions with which Picard was concerned are defined at all points of $M(F)$ and can be written in U_α as $R(x_\alpha, y_\alpha)$, a meromorphic function of (x_α, y_α) whose singular locus is on an algebraic curve of $M(F)$ and the simple and double integrals are integrals of forms ω^1, ω^2 defined globally on $M(F)$, where in U_α

$$\omega^1 = A(x_\alpha, y_\alpha)\,\mathrm{d}x_\alpha + B(x_\alpha, y_\alpha)\,\mathrm{d}y_\alpha,$$

$$\omega^2 = C(x_\alpha, y_\alpha)\,\mathrm{d}x_\alpha\,\mathrm{d}y_\alpha,$$

where A and B, C are meromorphic functions with singularities on an algebraic curve of $M(F)$. A 1-form ω^1 has associated uniquely with it a 2-form $\mathrm{d}\omega^1$, where locally,

$$\mathrm{d}\omega^1 = \left(\frac{\partial B}{\partial x_\alpha} - \frac{\partial A}{\partial y_\alpha}\right)\mathrm{d}x_\alpha\,\mathrm{d}y_\alpha,$$

and ω^1 is said to be closed if $\mathrm{d}\omega^1 = 0$.

The main problem investigated by Picard concerned the classification of integrals or differential forms according to the type of singularity they possessed. If ω^1 is closed 1-form on $M(F)$ a singularity on a curve C is said to be polar if, in the neighbourhood U_α of any point of C, there exists a meromorphic function R such that $\omega^1 - \mathrm{d}R$ is holomorphic; otherwise the singularity is logarithmic. A closed form ω^1 is of the first kind if it has no singularities, and of the second kind if all its singularities are polar. Similarly, for the singularities of 2-forms ω^2. A singular locus C of ω^2 is polar if in the neighbourhood of any part of it (lying, say, in U_α) there exists a 1-form $v^1 = A\,\mathrm{d}x_\alpha + B\,\mathrm{d}y_\beta$ such that $\omega^2 - \mathrm{d}v^1$ has no singularities in U_α. A global 2-form is of the first kind if it has no singular locus, and of the second kind if all its singularities are polar. Finally, two 1-forms ω^1 and v^1 are said to be *equivalent* if $\omega^1 - v^1 = \mathrm{d}R$, where R is a globally defined meromorphic function on $M(F)$, and two 2-forms ω^2 and v^2 are equivalent if there exists a globally defined 1-formed μ such that $\omega^2 - v^2 = \mathrm{d}\mu$.

Using only the most primitive notions of topology available to him, Picard succeeded in solving the classification problem completely by analytic means, but his reasoning is both very long and very involved. In the course of it, however, he did obtain what amounts to a detailed knowledge of the topology of $M(F)$, though the topological interpretation of some of his results was incomplete. It was at this point that Lefschetz made his great contribution. He recognized more fully than Picard had done the nature of the topology of $M(F)$, and he was able to describe the topology of an algebraic surface without having to use the theory of functions at all. He was then able to use the topology in the study of functions and forms on $M(F)$, so obtaining Picard's results in a form which yielded more information, and which made the extension of the results to varieties of more than two dimensions possible.

In particular, his investigation of the 1-cycles of $M(F)$ led to a simple account of simple integrals of the first kind in M. He first showed that the 1-cycles of $M(F)$ are homologous to cycles in a plane section of $M(F)$. By considering the automorphisms of the homology group of a section as the section varies he arrived at a subgroup (the group of invariant cycles) which forms a basis for the R_1 1-cycles of $M(F)$. Then from the invariance of these cycles he showed that $R_1 = 2q$. From this, it is a short step to show that there are q independent closed simple integrals of the first kind on $M(F)$, and $2q$ integrals of the second kind. Picard had found a similar result, but did not connect q with the topology of $M(F)$; he obtained q as the geometrical invariant known as the irregularity.

Lefschetz's account of the properties of double integrals follows in the main the work of Picard, but a knowledge of the topology of M makes the work much easier. A central problem concerns the period cycles of the double integrals, the difficulty arising from the fact that a 2-cycle of $M(F)$ in general meets a singular locus of the integral. A first step is to show that every double integral of the second kind is equivalent to one whose singularities are all on a fixed curve (the section C_0 of the surface by the plane at infinity). One can then discuss periods on the R_2-1 independent 2-cycles which do not meet the 2-cycle Γ_0 carried by C_0 (R_2 being the second Betti number of M). But although it is possible to construct integrals of the second kind having arbitrary periods on these cycles, some of these may themselves be equivalent to zero (in the sense defined above). The key result is that there exist $\rho-1$ curves $C_1, \ldots, C_{\rho-1}$ on M carrying cycles $\Gamma_1, \ldots, \Gamma_{\rho-1}$ with the property that a necessary and sufficient condition that an integral of the second kind be equivalent to zero is that it has zero period on every 2-cycle of $M(F)$ having zero intersection with $\Gamma_0, \ldots, \Gamma_{\rho-1}$. We then see that there are $R_2-\rho$ independent integrals of the second kind on M.

The set of curves $C_0, \ldots, C_{\rho-1}$ on F occurs in other connections; in particular, they form a base in the sense of Severi. In connection with these curves Lefschetz made a major contribution. Using a theory due to Poincaré, he showed that a necessary and sufficient condition that a 2-cycle Γ be homologous to the cycle carried by a curve is that every double integral of the first kind on M should have zero period on Γ. Thus the cycles carried by the curves of a Severi base form the sub-group of 2-cycles on which the integrals of the first kind have zero periods.

Although Lefschetz's main contribution to the general transcendental theory of varieties concerned algebraic surfaces, he made important contributions to the study of varieties of higher dimension, in particular in **[24]**, **[29]**, and **[30]**.

Special mention should be made of his work on Abelian varieties. An Abelian variety of p-dimensions is a locus V in N-space $(x_1, \ldots, x_N)$ defined by equation

$$x_i = \theta_i(u_1, \ldots, u_p) \quad (i = 1, \ldots, N),$$

where the θ_i are multiply periodic functions of p variables $(u_1, \ldots, u_p)$ having $2p$ periods, with a periodic matrix Ω of p rows and $2p$ columns. The case $p = 1$ is of course just the elliptic curve. For general p, the Riemannian manifold of V is the direct product of $2p$ circles, $\gamma_1, \ldots, \gamma_{2p}$ and the q-forms of the first kind on it form a vector space of ${}_pC_q$ dimensions, ${}_pC_q$ being the coefficient of x^q in $(1+x)^p$, a base for which consists of the forms $du_i, \ldots, du_{iq}$. The field is wide open for the exploitation of Lefschetz's topological methods, and 120 pages of his Prix Bordin memoir is devoted to this. We need only quote one property of Abelian varieties that follows most elegantly by Lefschetz's techniques. Since the variety is algebraic it contains a curve, which defines a 2-cycle, say, $\sum A_{ij}\gamma_i \times \gamma_j$, where $A = (A_{ij})$ is a skew-symmetric matrix of integers. The fact that every double integral of the first kind has zero period on the cycle yields the matrix equation $\Omega A \Omega' = 0$, and a variation of this argument leads to the inequality $i\Omega A\bar{\Omega}' > 0$. The existence of a matrix A associated with the period. matrix in this way is fundamental in the theory of multiply periodic functions Lefschetz's techniques lead to a complete theory of all the matrixes A which can be associated with a given periodic matrix Ω in this way.

It may be remarked here that many of Lefschetz's ideas in the theory of algebraic varieties over the complex field have been applied in a suitably generalized form to make possible some remarkable advances in number theory, when this is treated by geometrical methods. Perhaps the most striking example concerns Weil's conjectures for surfaces over finite fields and in particular the recent work of Deligne.

The period 1923 to about 1928 marks the shift of Lefschetz's interest from algebraic geometry to topology. The bridge was his concern with fixed-point theory, and his recognition that to exploit his ideas fully a rigorous theory of intersections on a topological manifold had to be developed. It will be convenient to leave discussion of the intersection theory to the next section of this memoir, and to describe here its application to geometrical problems including fixed-point problems. The extension of his methods to fixed-point problems on a more general manifold than the Riemannian of an algebraic variety will then be obvious.

The basic idea of Lefschetz's fixed-point theory is due to the Italian school of geometers, who generalized the elementary theory of graphs to give an account of the theory of correspondences between algebraic varieties. If A and B are two algebraic varieties and an algebraic correspondence γ is given between points of A and B, we get a geometrical representation of this correspondence by constructing the product $A \times B$ and considering the aggregate Γ of points (a, b) for which a and b are corresponding points in γ. Γ is an algebraic sub-variety of $A \times B$ and the theory of correspondences between A and B is equivalent to the theory of sub-varieties of $A \times B$. A case of particular importance is that in which $A = B$; if δ is the identity correspondence, its graph Δ on $A \times B$ is the diagonal of $A \times B$, and is a sub-variety of $A \times B$ birationally equivalent to A. The intersection of Δ with the graph of a correspondence γ on A is the image on $A\times B$ of the locus of points on A which are self-corresponding in γ.

Lefschetz first considered the case in which A and B are curves of genus p and q respectively, and γ is an (α, β) correspondence between them. $A\times B$ is a surface, whose topology is determined by the topology of A and B [37]. If $a, a_i (i = 1, \ldots, 2p)$, A are an 0-cycle, a set of independent 1-cycles, and the carrier 2-cycle of A, and $b, b_i (i = 1, \ldots, 2q)$, B are similarly defined for B, $a\times B$, $a_i\times b_j (i = 1, \ldots, 2p, j = 1, \ldots, 2q)$, $A\times b$ form a base for the 2-cycles of $A\times B$. The graph is an algebraic 2-cycle Γ homologous to

$$\beta a\times B+\alpha A\times b+\sum r_{ij}\, a_i\times b_j,$$

where the r_{ij} are integers. The double integrals of the second kind on $A\times B$ are easily determined from the simple integrals $u_1, \ldots, u_p$ and $v_1, \ldots, v_q$ on A and B respectively. If the period matrices of these are ω and v, the fact that the double integrals on $A\times B$ have zero periods on Γ leads to the matrix relation $\omega r v' = 0$. Conversely, if there does exist such a relation between ω and v, it follows from Lefschetz's general theorem about 2-cycles on a surface on which the double integrals of the first kind have zero-periods that there exists an algebraic curve on $A\times B$ homologous, as a cycle, to

$$\beta a\times B+\alpha A\times b+\sum r_{ij}\, a_i\times b_j$$

for any integers α, β (and that, by choosing these suitably, this curve is effective an irreducible). Γ thus defines an (α, β) correspondence between A and B.

Lefschetz exploited this method of studying the correspondences between curves fully. The case in which $A = B$ is of particular importance. We may assume that $b_i = a_i$, and denote by m_{ij} the intersection number of a_i, a_j on A. If Δ is the diagonal of $A\times B$, Lefschetz showed that as a 2-cycle it was homologous to

$$a\times B+A\times b+\sum \varepsilon_{ij}\, a_i\times b_j$$

where (ε_{ij}) is the transpose of the inverse of (m_{ij}). An immediate application of the result quoted above gives the Riemann equalities $\omega\varepsilon\omega' = 0$, and an argument on the same lines leads to the Riemann inequalities. If t is a rational transformation of A into itself (i.e. a $(\alpha, 1)$ correspondence) it is represented by its graph on $A\times B$ and is an algebraic cycle

$$T = a\times B+\alpha A\times b+\sum r_{ij}\, a_i\times b_j.$$

The coefficients r_{ij} are determined by the action of t on the first homology group of A. The intersection number $\Gamma.\ \Delta$ is equal to $1+\alpha+\sum r_{ij} m_{ij}$, and this gives the number of fixed points of the transformation t.

This method of determining the number of fixed points of t is purely topological, and does not depend on properties of the integrals on A or $A\times B$. Hence it can be extended to the case in which A is any orientable manifold, and T is a continuous transformation of A into itself. The only new point to be noticed is that the points common to Γ and Δ have a sign attached to them, and the intersection number $\Gamma.\Delta$

is the sum of the signed intersection numbers. This number $\Gamma.\Delta = L(T)$ is computed to be

$$L(T) = \sum_{r=0}^{n} (-1)^r \text{ trace } T_r$$

where T_r is the operation on the rth homology group $H_r(A)$ (rational coefficients) of A, a formula first given in [25].

While Lefschetz exploited his graphical method of dealing with correspondences fully in the case of curves, he did not pursue it as much as he might have done for algebraic varieties of higher dimensions. This may have been due to the fact that there was no condition available for determining whether a $2r$-cycle on a variety of n dimensions belonged to the homology class defined by a sub-variety of complex dimension r, except when $r = 1$, $n = 2$, but one might hazard the guess that it was because he saw the significance of his methods in a purely topological situation and directed his attention to generalizations in this field with his usual single-mindedness.

(*b*) *Topology*

From the time of his appointment to the Faculty at Princeton in 1924 until 1938, Lefschetz's researches were very largely confined to topology. This period was in marked contrast with his time in Nebraska and Kansas, since a very great deal of his work in algebraic geometry was done in isolation, whereas his work in topology was done as a member of a flourishing school whose work was world-famous. The consequence was that although his personal contributions, particularly to fixed-point theory, duality, and intersections, are of the first magnitude, a great deal of his efforts went into setting out a comprehensive account of topological theory, as in his Colloquium Lectures (45, 79), and in his inspiration of the younger generation. In this he was so successful that it is no surprise to find so many of his pupils in leading positions in the mathematical world.

Lefschetz's entry into the field of algebraic topology (he was in fact influential in introducing the modern use of this term) began as explained above, with the graphical representation of correspondences between algebraic varieties, which he took from the Italian geometers, but at the same time he was aware of the theorems of L. E. J. Brouwer on continuous mappings of an n-cell or an n-sphere into itself, and also of a theorem of J. W. Alexander on topological mappings of a two-dimensional manifold. What Lefschetz did was to put these earlier ideas together and establish a general form of these early results, and his genius was displayed in the way in which he unerringly selected the " right " generalization, and in the thoroughness and ingenuity with which he carried through the details of such a generalization.

The essential steps in replacing the geometrical problem of fixed-points by a topological one are the recognition (1) that a non-singular algebraic variety V of dimension m over the complex field is an orientable topological manifold $M(V)$ of dimension $2m$, and (2) that any sub-variety of V of complex dimension r defines a $2r$-cycle of $M(V)$. It has then to be shown that if Γ and Γ' are any two cycles of dimension s and t of $M(V)$ they can be replaced by homologous cycles Γ and Γ' whose intersection is a cycle of dimension $2m-s-t$, whose homology class is completely determined by Γ and Γ'. When Γ and Γ' are the cycles carried by sub-varieties U and U' of complex dimension ρ, σ which meet simply in a sub-variety of complex dimension $\rho+\sigma-m$, the intersection $\Gamma.\Gamma'$ is homologous to the carrier of $U.U'$. In particular, when U, U' are effective algebraic sub-varieties, of dimensions $\rho, m-\rho$, the intersection number $(\Gamma.\Gamma')$ is the actual number of points common to U and U', each counted with its proper multiplicity.

The topological theory of intersection applies to any manifold M of dimension n. The fact that M is a manifold amounts to saying that it can be covered by a polyhedral complex K with the property that if $E_p{}^i$ is any p-cell of K, the aggregate of $E_p{}^i$ and all the cells of K which have $E_p{}^i$ on their face is an n-cell (the star of $E_p{}^i$). We construct a barycentric sub-division K' of K by introducing into each cell

$$E_p{}^i(i = 1, 2, \ldots, p = 1, \ldots, n)$$

a new vertex $A_p{}^i$, and joining $A_1{}^i$ to the 0-cells of K on the boundary of $E_1{}^i$, $A_2{}^i$ to the 1-cells so constructed on the boundary of $E_2{}^i$, and so on. Then if we consider for any $A_p{}^i$ the aggregate of all the cells $(A_p{}^i A_{p1}{}^{i_1} \ldots A_{pa}{}^{ia})$ when $p < p_1 < \ldots < p_a \leqslant n$ so constructed, it is shown from the manifold condition that these constitute a regular sub-division of an $(n-p)$-cell E^{*i}_{n-p} lying in the star of $E_p{}^i$. These cells E^{*i}_{n-p} (each with an arbitrarily assigned orientation) constitute a dual complex K^* which also covers M. It can be shown that the cells $E_p{}^i$ and $E_q{}^{*j}$ either have no point in common or meet in a cell of dimension $p+q-n$. To construct an intersection theory on M we use the fact and any r-cycle of M can be replaced by an r-cycle Γ of K, and any s-cycle by a cycle Γ' of K^*. Then as a point set the points common to Γ and Γ' is an aggregate of cells in K. But to get a useful topological notion of intersection one must introduce the notion of the oriented intersection. What Lefschetz did first was to assume that M is orientable. Then it is possible to give an intrinsic definition of the oriented dual of E^{*i}_{n-p} of $E_p{}^i$ in terms of the orientations of $E_p{}^i$ and of M and then define the oriented cell $E_p{}^i.E_q{}^{*j}$. It is then shown that if $\Gamma = \sum a_i E_p{}^i$, $\Gamma' = \sum b_j E_q{}^{*j}$, $\Gamma.\Gamma' = \sum a_i b_j E_p{}^i.E_q{}^{*j}$ is a cycle of K' whose homology class is completely determined by the homology classes of Γ and Γ'. Of particular importance is the case in which $q = n-p$. Then the conventions regarding orientation lead to $E_p{}^i.E^{*l}_{n-p} = \delta^{lj} A_p{}^i$. The number $\sum_i a_i b_i$ is then the number of signed intersections of Γ and Γ'.

As with all his contributions to mathematics, Lefschetz's ideas on intersections sprang from an intuitive geometrical conception of the problem which he gradually developed in a series of papers, particularly [31], [32], [33], until he had a precision instrument at his disposal. But his applications of intersections did not wait until his intersection theory was complete: thus while his main work on intersections was done in 1925–26 he had already, in 1923, used his more primitive notions of intersections to obtain his fixed-point formula for continuous self-transformations of an orientable manifold [25].

Lefschetz was not, however, satisfied to give this formula only for absolute orientable manifolds, and most of his later work in topology was concerned with the study of more general spaces for which he could obtain some kind of fixed-point theorem. First, consider a manifold which is not orientable. Then if K is the covering complex there is no preferred way of orientating the $(n-p)$-cell E^{*i}_{n-p} dual to the oriented cell $E_p{}^i$. Dual cells intersect in an 0-cell, but there is no way of choosing between the two orientations of this cell. The consequence is that we can only show that the intersection of the graph of the correspondence T and the diagonal cycle of $M \times M$ is a set of unoriented 0-cells. The parity of the number of 0-cells is equal to $L(T)$ reduced modulo 2, and the only conclusion we can draw is that if $L(T)$ is odd, then T must have at least one fixed point.

Lefschetz's formula gave Brouwer's result for n-spheres, but failed to give the Brouwer theorem for n-cells. This is because an n-cell is not a manifold, since it has a boundary. To cover the Brouwer theorem and give a generalized version of it, Lefschetz considered relative manifolds. A relative manifold M of n dimensions is a complex K, with a sub-complex L, such that (*a*) L is a manifold, and (*b*) every cell of $K-L$ has the property that its star is an n-cell. Then we can, as in the absolute case, construct an $(n-p)$-cell E^{*i}_{n-p} dual to each cell $E_p{}^i$ of $K-L$. It is possible to construct a theory of cycles of K modulo L in which any chain whose boundary lies in L is a relative cycle, and a theory of cycles formed from the cells $E_q{}^{*i}$ which lie in $K-L$. There then exists the same type of duality (omitting torsion) between the groups $H_p(K, \text{mod } L)$ and the group $H_{n-p}(K-L)$, which is the group of $(n-p)$-cycles of $K-L$ " with compact supports ". Lefschetz's contribution to duality theory, for which Poincaré and Alexander had laid the basis, was a major contribution. It also allows the development of fixed-point theory for relative manifolds.

In the years 1936–39, Alexander, Cech and Whitney developed the theory of cohomology, which led Lefschetz to re-cast his intersection theory [79]. In this theory a p-cochain $\mathscr{C}^p$ on a manifold M is a linear functional of the p-chains $\mathscr{C}_p$ of the space, with a coboundary operator d which defines $d\mathscr{C}^p$ as the $(p+1)$-cochain defined by $d\mathscr{C}^p(C_{p+1}) = \mathscr{C}^p(\partial C_{p+1})$. The group of p-cocycles (cochains $\mathscr{C}^p$ such that $d\mathscr{C}^p = 0$) reduced modulo the p-cocycles which are coboundaries, is the cohomology group $H^p(M)$. On an orientable manifold M covered by a complex K we show that each p-cocycle $\mathscr{C}^p$ can be represented as an $(n-p)$-cycle C^*_{n-p} of the dual complex K^*, and define the value of $\mathscr{C}^p$ on C_p as the intersection number $C_p . C^*_{n-p}$. In this way there is an isomorphism $H^p(M) \sim H_{n-p}(M)$. Much flexibility is gained by replacing p-cocycles by their corresponding $(n-p)$ cycles. Then we have to consider the $(p+q)$-cocycle which is dual to the cycle $\Gamma_{n-p}\,\Gamma_{n-q}$ where Γ_{n-p}, Γ_{n-q} are dual to the cocycles Γ^p, Γ^q. In this way the basic notion of the cup-product of cocycles is obtained. Lefschetz's version of cohomology theory, with cup-products, did not appear in full until 1942 [79], but it is characteristic of him that as early as 1930 he was feeling his way to the general theory with his notion of pseudo-cycles [**48**].

Finally, we must mention Lefschetz's contribution to the theory of local connectedness. The problem here is to find types of space more general than manifolds in which one can construct a homology theory with duality theorems. P. Alexandroff, Cech and others investigated this problem in depth. Lefschetz's contribution [**55**, **64**] dealt with local connectedness of a space X. X is said to be locally connected (in the sense of homology), HLC_q in dimension q, if it has the property that given any point x of X, and any neighbourhood V of x, then there is a neighbourhood U_q of x such that any continuous image of a q-sphere in U_q bounds in V. If X is HLC_q for $q = 0, 1, 2, \ldots$ and for any x and V we can choose the same $U_q = U$ for all q, X is HL C. He showed that the fixed-point theorem is valid for compact HLC spaces.

(*c*) *Differential equations*

Lefschetz virtually ceased making original contributions to algebraic geometry and topology after the publication of his *Algebraic topology* [**79**] in 1942, though he continued to lecture and write books and expository articles on these subjects for many years after that. But his main work for the last 25 years of his life was on the subject of differential equations and their applications. A number of circumstances combined to produce this major change in the role he played in mathematics. He was nearly 60 years old, an age by which most research mathematicians are prepared to call it a day; and he had already squeezed as much out of the important ideas which he had introduced into geometry and topology as he was temperamentally capable of. Further, the United States was now in World War II, during most of which he served as a consultant to the Department of the Navy. It was during this period that he was introduced by Nicholas Minorsky to the work done by Russian mathematicians in nonlinear oscillations and stability. He immediately realized the importance of the geometric theory of differential equations which had grown out of the work of Poincaré and Liapunov. Feeling that this subject has been far too long neglected he set about remedying the situation with all his usual vigour and enthusiasm. His energy was such that in spite of his years he was able quickly to gather round him an active band of workers and for the rest of his life he was the leader of a vigorous and distinguished school (as described in the first part of this memoir). By making available translations of the leading Russian workers in the field, by lecturing, by writing textbooks and review articles, as well as by original papers of his won, he stirred up enthusiasm and created one of the leading schools in the country. Many younger men owe their introduction to the subject to his books.

It was not to be expected that at his age he could bring the same wealth of original ideas to differential equations and control theory that he had earlier brought to geometry and topology. Nevertheless he made important original contributions, a few of which may be noted.

Most noteworthy of his original papers are the ones concerned with the behaviour of the solutions of analytic differential equations near an isolated singular point. In this area, he brought to bear his tremendous knowledge of the theory of algebraic curves. In paper [**129**] he gave a complete characterization and a constructive procedure for obtaining all the solution curves of a two-dimensional system near an isolated critical point which pass through this critical point. Relating this to the complete phase portrait, he made a significant improvement in the original work of Bendixson.

For a two-dimensional analytic system for which the coefficient matrix of the linear variational equation of an isolated critical point has both roots zero but is not identically zero, he showed (in [**106**]) that there can be at most a single nested oval of orbits.

Paper [**111**] is still one of the best treatments of the manner in which one determines stability of an isolated equilibrium point of an n-dimensional system for which the linear variational equation has some zero roots. In papers [**83**] and [**97**], Lefschetz also considered the existence of periodic solutions of second and higher order nonlinear systems of differential equations.

I wish to acknowledge my indebtedness for the help I have received in writing this memoir from many sources. Professor W. Browder of Princeton University has supplied me with much useful information regarding the biographical details, besides sending me copies of numerous notices which appeared in America. He also obtained the portrait for me. The appreciation of Professor Lefschetz's work in algebraic geometry and topology is based on two articles, respectively by myself and the late Professor N. E. Steenrod, which were written for the symposium held in 1954 in honour of his seventieth birthday, and published by the Princeton University Press in 1957. Finally, Professors J. K. Hale and J. P. LaSalle of Brown University have allowed me to make use of an article which they wrote on the life and work of Professor Lefschetz during the last 25 years of his life.

References

1912 **1.** " Two theorems on conics ", *Ann. Math.*, (2) 14, 47–50.

2. " On the V_3^3 with five nodes of the second species in S_4 ", *Bull. Am. Math. Soc.*, 18, 384–386.

3. " Double curves of surfaces projected from space of four dimensions ", *Bull. Am. Math. Soc.*, 19, 70–74.

1913 **4.** " On the existence of loci with given singularities ", *Trans. Am. Math. Soc.*, 14, 23–41. Doctoral dissertation, Clark University, 1911.

5. " On some topological properties of plane curves and a theorem of Möbius ", *Am. J. Math.*, 35, 189–200.

1914 **6.** " Geometry on ruled surfaces ", *Am. J. Math.*, 36, 392–394.

7. " On cubic surfaces and their nodes ", *Kansas Univ. Science Bull.*, 9, 69–78.

1915 **8.** " The equation of Picard–Fuchs for an algebraic surface with arbitrary singularities, *Bull. Am. Math. Soc.*, 21, 227–232.

9. " Note on the n-dimensional cycles of an algebraic n-dimensional variety ", *R. C. Mat. Palermo*, 40, 38–43.

1916 **10.** " The arithmetic genus of an algebraic manifold immersed in another ", *Ann. Math.* (2), 17, 197–212.

11. Direct proof of De Moivre's formula ", *Am. Math. Monthly*, 23, 366–368.

12. " On the residues of double integrals belonging to an algebraic surface ", *Quart. J. Pure Appl. Math*, 47, 333–343.

1917 **13.** " Note on a problem in the theory of algebraic manifolds ", *Kansas Univ. Science Bull.*, 10, 3–9.

14. " Sur certains cycles à deux dimenions des surfaces algébriques ", *R. C. Accad. Lincei* (5), 26, 228–234.

15. " Sur les intégrales multiples des variétés algébriques ", *C. R. Acad. Sci. Paris*, 164, 850–853.

16. " Sur les intégrales doubles des variétés algébriques. *Annali Mat.* (3), 26, 227–260.

1919 **17.** " Sur l'analyse situs des variétés algébriques ", *C, R. Accad. Scri. Paris*, 168, 672–674.

18. " Sur les variétés abéliennes ", *C. R. Accad. Sci. Paris*, 168, 758–761.

19. " On the real folds of Abelian varieties ", *Proc. natn. Acad. Sci U.S.A.*, 5, 103–106.

20. " Real hypersurfaces contained in Abelian varieties ", *Proc. Natn. Acad. Sci. U.S.A.*, 5, 296–298.

1920 **21.** " Algebraic surfaces, their cycles and integrals ", *Ann. Math.* (2), 21, 225–228. A correction, *ibid*, 23, 333.

1921 **22.** " Quelques remarques sur la multiplication complexe ", *Comptes Rendus du Congrés International des Mathématiciens, Strasbourg*, September 1920, Toulouse: E. Privat.

23. " Sur le théorème d'existence des fonctions abéliennes ", *R. C. Accad. Lincei*, (5), 30, 4:–50.

24. " On certain numerical invariants of algebraic varieities with application to Abelian varieties ", *Trans. Am. Math. Soc.*, **22**, 327–482.

1923 **25.** " Continuous transformations of manifolds ", *Proc. natn. Acad. Sci. U.S.A.* 9, 90–93.

26. Progres récents dans la théorie des fonctions abeliennes ", *Bull. Sci. Math.* **47**, 120–128.

27. " Sur les intégrales de seconde espèce des variétés algébriques ", *C. R. Acad. Sci. Paris*, 176, 941–933.

28. " Report on curves traced on algebraic surfaces ", *Bull. Am. Math. Soc.*, 29, 242–258.

1924 **29.** *L'analysis situs et la géométrie algébrique* (Paris: Gauthier-Villars). (Collection de Monographies publiée sous la direction de M. Emile Borel.) Nouveau tirage, 1950.

30. " Sur les intégrales multiples des variétés algébriques ", *J. Math. Pures Appl.*, (9), 3, 319–343.

1925 **31.** " Intersections of complexes on manifolds ", *Proc. natn. Acad. Sci. U.S.A.*, 11, 287–289.

32. " Continuous transformations of manifolds ", *Proc. natn. Acad. Sci U.S.A.*, 11, 290–292

1926 **33.** " Intersections and transformations of complexes and manifolds ", *Trans. Am. tath.. Soc.*, 28, 1–49.

34. " Transformations of manifolds with a boundary ", *Proc. natn. Acad. Sci. U.S.A.*, 12, 737–739.

1927 **35.** " Un théorème sur les fonctions abéliennes ", *In Memoriam N. I. Lobatschevskii*, pp. 186–190. Kazan's, Glavnauka.

36. " Manifolds with a boundary and their transformations ", *Trans. Am. Math. Soc.*, 29 429–462, 848.

37. " Correspondences between algebraic curves ", *Ann, Math.* (2), 28, 342–354.

38. " The residual set of a complex on a manifold and related questions ", *Proc. natn. Acad. Sci. U.S.A.*, 13, 614–622, 805–807.

39. " On the functional independence of ratios of theta functions ", *Proc. natn. Acad. Sci. U.S.A.*, 13, 657–659.

1928 **40.** " Transcendental theory; Singular correspondences between algebraic curves; Hyperelliptic surfaces and Abelian varieties ", in *Selected topics in algebraic geometry*, vol. 1, chap. 15–17, pp. 310–395. Report of the Committee on Rational Transformations of the National Research Council, Washington (N.C.R. Bulletin no. 63).

41. " A theorem on correspondences on algebraic curves ", *Am. J. Math.*, 50, 159–166.

42. " Closed point sets on a manifold ", *Ann. Math.*, (2), 29, 232–254.

1929 **43.** *Géométrie sur les surfaces et les variétés algébriques* (Paris: Gauthier-Villars). (Mémorial des Sciences Mathématiques, Fasc. 40.)

44. " Duality relations in topology ", *Proc. natn. Acad. Sci. U.S.A.*, 15, 367–369.

1930 **45.** *Topology* (New York: American Mathematical Society). (Colloquium Publications, vol. 12.)

46. " Les transformations continues des ensembles fermés et leurs points fixes ", *C. R. Acad. Sci. Paris*, 190, 99–100.

47. (With W. W. Flexner) " On the duality theorems for the Betti numbers of topological manifolds ", *Proc. natn. Acad. Sci. U.S.A.*, 16, 530–533.

48. " On transformations of closed sets ", *Ann. Math.* (2), 31, 271–280.

1931 **49.** " On compact spaces ", *Ann. Math.*, (2), 32, 521–538.

1932 **50.** " On certain properties of separable spaces ", *Proc. natn. Acad. Sci. U.S.A.*, 18, 202–203.

51. " On separable spaces ", *Ann. Math.*, (2), 33, 525–537.

52. " Invariance absolute et invariance relative en géométrie algébrique ", *Rec. Math.* (*Mat. Sbornik*), 39, 97–102.

1933 **53.** " On singular chains and cycles ", *Bull. Am. Math. Soc.*, 39, 124–129.

54. (With J. H. C. Whitehead) " On analytical complexes ", *Trans. Am. Math. Soc.*, 35, 510–517.

55. " On generalized manifolds ", *Am. J. Math.*, 55, 469–504.

1934 **56.** *Elementary one- and two-dimensional topology* (Princeton University (mimeographed)).

57. " On locally connected and related sets ", *Ann. Math.* (2), 35, 118–129.

1935 **58.** *Topology* (Princeton University (mimeographed)).

59. " Algebraicheskaia geometriia: metody, problemy, tendentsii ", in *Trudy Vtorogo Vsesoiuznogo Matematischeskogo S " ezda, Leningrad*, 24–30 *June* 1934, vol. 1, pp. 337–349. Leningrad–Moscow.

60. " Chain-deformations in topology ", *Duke Math. J.*, 1, 1–18.

61. " Application of chain-deformations to critical points and extremals ", *Proc. natn. Acad. Sci. U.S.A.*, 21, 220–222.

62. " A theorem on extremals. I, II ", *Proc. natn. Acad. Sci. U.S.A.*, 21, 272–274, 362–364.

63. " On critical sets ", *Duke Math. J.*, 1, 392–412.

1936 **64.** " On locally-connected and related sets (second paper) ", *Duke Math. J.*, 2, 435–442.

65. " Locally connected sets and their applications ", *Rec. Math.* (*Mat. Sbornik*), n.s., 1, 715–717.

66. " Sur les transformations des complexes en sphères ", *Fund. Math.*, 27, 94–115.

67. Matematicheskaia deiatel'nost' v Prinstone ", *Uspekhi Mat. Nauk*, 1, 271–273.

1937 **68.** *Lectures on algebraic geometry* 1936–37 (Princeton University (planographed)).

69. " Algebraicheskaia geometriia ", *Uspekhi Mat*, 3, 63–77.

70. " The role of algebra in topology ", *Bull. Am. Math. Soc.*, 43, 345–359.

71. " On the fixed point formula ", *Ann. Math.*, (2), 38, 819–822.

1938 **72.** *Lectures on algebraic geometry* (part II) 1937–1938 (Princeton University Press (planographed)).

73. " On chains of topological spaces ", *Ann. Math.* (2), 39, 383–396.

74. " On locally connected sets and retracts ", *Proc. natn. Acad. Sci. U.S.A.* 24, 392–393.

75. " Sur les transformations des complexes en sphères (note complémentaire) ", *Fund. Math.*, 31, 4–14.

76. " Singular and continuous complexes, chains and cycles ", *Rec. Math.* (*Mat. Sbornik*), n.s., 3, 271–285.

1939 **77.** " On the mapping of abstract spaces on polytopes ", *Proc. natn. Acad. Sci. U.S.A.*, 25, 49–50.

1941 **78.** " Abstract complexes ", in *Lectures in topology; the University of Michigan Conference of* 1940, pp. 1–28 (University of Michigan Press).

1942 **79.** *Algebraic topology* (New York, American Mathematical Society). (Colloquium Publications, vol. 27.)

80. *Topics in topology* (Princeton University Press). (Annals of Mathematics Studies, no. 10.) A second printing, 1951.

81. Emile Picard (1856–1941): Obituary. *Am. Phil. Soc. Yearbook* 1942, pp. 363–365.

1943 **82.** *Introduction to non-linear mechanics*, by N. Kryloff and N. Bogoliuboff (a free translation by S. Lefschetz) (Princeton University Press). (Annals of Mathematics Studies, no. 11.)

83. " Existence of periodic solutions for certain differential equations ", *Proc. natn. Acad. Sci. U.S.A.*, 29, 29–32.

1946 **84.** *Lectures on differential equations* (Princeton University Press). (Annals of Mathematics Studies, no. 14.)

1949 **85.** *Introduction to topology* (Princeton University Press). (Princeton Mathematical Series, no. 11.)

86. *Theory of oscillations*, by A. A. Andronow and C. E. Chaikin (English language edition edited under the direction of S. Lefschetz) (Princeton University Press).

87. " Scientific research in the U.S.S.R.: Mathematics ", *Am. Acad. Polit. and Soc. Sci. Annals*, 263, 139–140.

1950 **88.** *Contributions to the theory of nonlinear oscillations* (edited by S. Lefschetz) (Princeton University Press). (Annals of Mathematics Studies, no. 20.)

89. " The structure of mathematics ", *American Scientist*, 38, 105–111.

1951 **90.** " Numerical calculations in nonlinear mechanics ", in *Problems for the numerical analysis of the future*, pp. 10–12 (Washington: Govt. Printing Office). (National Bureau of Standards, Applied Math. Series, no. 15.)

1952 **91.** *Contributions to the theory of nonlinear oscillations* (edited by S. Lefschetz), vol. 2 (Princeton University Press). (Annals of Mathematics Studies, no. 29.)

92. " Notes on differential equations ", in *Contributions to the theory of nonlinear oscillations*, vol. 2, pp. 67–73.

1953 **93.** *Algebraic geometry* (Princeton University Press). (Princeton Mathematical Series, no. 18.)

94. " Algunos trabajos recientes sobre ecuaciones diferenciales ", in *Memoria de Congreso Cientifico Mexicano, U.N.A.M., Mexico*, vol. 1, pp. 122–123.

95. " Las grades corrientes en las matematicas del siglo XX ", in *Memoria de Congreso Cientifico Mexicano, U.N.A.M., Mexico*, vol. 1, pp. 206–211.

1954 **96.** " Russian contributions to differential equations ", in *Proceedings of the Symposium on Nonlinear Circuit Analysis, New York*, 1953, pp. 68–74 (New York, Polytechnic Institute Brooklyn).

97. " Complete families of periodic solutions of differential equations ", *Comment Math. Helv.*, 28, 341–345.

98. " On Liénard's differential equation ", in *Wave motion and vibration theory*, pp. 149–153 (New York, McGraw-Hill). (Am. Math. Soc. Proceedings of Symposia in Applied Math., vol. 5.)

1956 **99.** " On a theorem of Bendixson ", *Bol. Soc. Mat. Mexicana*, 1, 13–27.

100. *Topology*, 2nd ed. (New York, Chelsea Publishing Company). (Cf. (45) above.)

1957 **101.** " On coincidences of transformations ", *Bol. Soc. Mat. Mexicana*, 2, 16–25.

102. " The ambiguous case in planar differential systems ", *Bol. Soc. Mat. Mexicana*, 2, 63–74.

103. Withold Hurewicz: in memoriam, *Bull. Am. Math. Soc.*, 63, 77–82.

104. " Sobre la modernizacion de la geometria ", *Revista Matematica*, 1, 1–11.

105. *Differential equations: geometric theory* (New York: Interscience). (Cf. (116) below.)

1958 **106.** " On the critical points of a class of differential equations ", in *Contributions to the theory of non-linear oscillations*, vol. 4, pp. 19–28 (Princeton University Press).

107. " Liapunov and stability in dynamical systems ", *Bol. Soc. Mat. Mexicana* (2), 3, 25–39.

108. *The Stability Theory of Liapunov* (University Institute for Fluid Dynamics and Applied Mathematics, College Park, Md.). (Lecture Series no. 37.)

1960 **109.** " Controls: an application to the direct method of Liapunov ", *Bol. Soc. Mat. Mexicana*, 5, 139–143.

110. " Algunas consideraciones sobre las matematicas modernas ", *Revista Union Matematica Argentina*, 20, 7–16.

111. " Resultados nuevos sobre casos criticos en ecuaciones diferenciales ", *Revista Union Matematica*, 20, 122–124.

1961 **112.** " The critical case in differential equations ", *Bol. Soc. Mat. Mexicana*, 6, 5–18.

113. *Geometricheskaia teoriia differentsial'nykh uravnenii* (Moskva, Izd-vo Inostrannoi Lit-ry). (Translation of 105)).

114. (With J. P. Lasalle) *Stability by Liapunov's direct method* (New York: Academic Press).

115. (With J. P. Lasalle) " Recent Soviet contributions to ordinary differential equations and non-linear mechanics ", *J. Math. Analysis Appl.*, 2, 467–499.

1962 **116.** *Differential equations: geometric theory* (2nd (rev.) ed.) (New York: Interscience).

117. *Recent Soviet contributions to mathematics* (edited with J. P. Lasalle) (New York: Macmillan).

1963 **118.** " On indirect automatic controls ", *Trudy Mezhdunarodnogo Simpoziuma po Nelineinym Kolebaniyam. Kiev, Izdat. Akad. Nauk Ukrain. SSR*, pp. 23–24.

119. " Some mathematical considerations on nonlinear automatic controls ", in *Contributions to differential equations*, vol. 1, pp. 1–28 (New York, Interscience).

120. *Elementos de toplogogia* (México, Universidad Naciónal Autonoma de México).

121. *Proceedings of international symposium on nonlinear differential equations and nonlinear mechanics, Colorado Springs*, 1961 (edited with J. P. Lasalle) (New York: Academic Press).

1964 **122.** *Stability of nonlinear automatic control systems* (New York: Academic Press).

1965 **123.** " Liapunov stability and controls ", *SIAM J. Control, Ser. A*, 3, 1–6.

124. " Planar graphs and related topics ", *Proc. natn. Acad. Sci. U.S.A.*, 54, 1763–1765.

125. " Recent advances in the stability of nonlinear controls ", *SIAM Rev.*, 7, 1–12.

126. " Some applications of topology to networks ", in *Proc. third annual Allerton conference on circuit and system theory*, pp. 1–6 (Urbana: University of Illinois).

1966 **127.** *Stability in dynamics* (Princeton: Princeton University School of Engineering and Applied Science). (William Pierson Field Engineering Lectures, 3, 4, 10, 11 March, 1966.)

1967 **128.** *Stability of nonlinear automatic control systems* (in Russian) (Moscow: Izdat. " Mir "). (A translation of (122).)

1968 **129.** " On a theorem of Bendixson ", *J. Diff. Equations*, 4, 66–101.

130. " A page of mathematical autobiography ", *Bull. Am. Math. Soc.*, 74, 854–879.

1969 **131.** " The Lurie problem on nonlinear controls ", in *Lectures in differential equations* (edited by A. K. Aziz) vol. 1, pp. 9–19 (New York: Van Nostrand–Reinhold).

132. " The early development of algebraic geometry ", *Am. Math. Monthly*, 76, 451–460.

133. " Luther Pfahler Eisenhart ", 1876–1905: A biographical memoir ", *Natn. Acad. Sci. U.S.A., Biographical Memoirs*, 40, 69–90.

1971 **134.** " The early development of algebraic topology ", *Bol. da Soc. Brasileira de Matematico*, 1, 1–48.

Contemporary Mathematics
Volume 58, Part I, 1986

MULTIPLICITY STRUCTURES ON SPACE CURVES

C. Bănică and O. Forster

Introduction. Let Y be an analytic (resp. algebraic) curve in a 3-dimensional complex analytic (resp. algebraic) manifold X. In several occasions one has to consider on Y not only the reduced structure, but a "multiplicity structure", which is defined by an ideal $J \subset \mathcal{O}_X$ with zero set V(J) = Y but which does not necessarily consist of all functions vanishing on Y. The structure sheaf $\mathcal{O}_X/J$ of the multiplicity structure may then contain nilpotent elements. For example let Y be a smooth (or more generally locally complete intersection) algebraic curve in affine 3-space $\mathbb{A}^3$. Ferrand/Szpiro (see [6]) have shown that Y is a set-theoretic complete intersection. The two polynomials f,g which describe Y set-theoretically generate an ideal J which defines a multiplicity 2 structure on Y. For the proof of this theorem, the ideal J is constructed first in such a way that the conormal module J/J^2 is globally free of rank 2 and then it follows from a theorem of Serre that J can be generated by 2 elements.

Another instance where curves with multiplicity structures are useful is in the study of vector bundles of rank 2 on 3-manifolds. Here the curves occur as zero sets of sections of the bundle. These curves carry a natural multiplicity structure. Under some hypotheses one can reconstruct the bundles from the curves (see e.g. [1],[2],[4],[5]).

In this paper, after introducing some notations and conventions, we recall first the Ferrand construction for multiplicity 2 structures and proceed then to a systematic study of structures of higher multiplicity, whose reduction is a smooth curve. Up to multiplicity 4 we obtain a complete descripton.

§ 0. Notations and generalities

(0.1) Although most of the results are also valid in the algebraic case, we work here in the analytic category. By a manifold we mean always a complex-analytic manifold X. An analytic subspace $Z \subset X$ may be non-reduced, i.e. is a pair $Z = (|Z|, \mathcal{O}_Z)$, where the structure sheaf is of the form $\mathcal{O}_Z = \mathcal{O}_X/I_Z$, where $I_Z \subset \mathcal{O}_X$ is a coherent ideal sheaf with zero-set $|Z|$. For two subspaces Z_1, Z_2 of X we write $Z_1 \subset Z_2$ if $I_{Z_1} \supset I_{Z_2}$. The intersection $Z_1 \cap Z_2$ is the subspace defined by the Ideal $I_{Z_1 \cap Z_2} := I_{Z_1} + I_{Z_2}$.

(0.2) In this paper we are mainly concerned with the following situation: There is given a smooth subspace (i.e. submanifold) $Y \subset X$ and another subspace $Z \supset Y$ of X with $|Z| = |Y|$. In a neighborhood of a point $a \in Y$ there exists a holomorphic retraction $X \longrightarrow Y$, hence also a retraction

$$\pi : Z \longrightarrow Y,$$

which is the identity on the underlying topological spaces. (More precisely, one should write $\pi : Z \cap U \longrightarrow Y \cap U$, U neighborhood of a. But we omit the indication of U for simplicity of notation.)

Now the following conditions are equivalent:

i) Z is Cohen-Macaulay (i.e. all local rings $\mathcal{O}_{Z,z}$ are Cohen-Macaulay).

ii) π is a flat map.

iii) The image sheaf $\pi_* \mathcal{O}_Z$ is locally free over $\mathcal{O}_Y$.

If Y is connected, the rank of $\pi_* \mathcal{O}_Z$ is then constant and equal to the multiplicity of Z.

If Z is Cohen-Macaulay, the multiplicity can be calculated also in the following way: In a neighborhood of a point $a \in Z$ let H be a submanifold of X with $\dim_a Y + \dim_a H = \dim_a X$ and such that H and Y intersect transversally at a. Then the multiplicity of Z at a equals

$$\mu = \dim_{\mathbb{C}} \mathcal{O}_{H \cap Z, a} .$$

(0.3) The intersection $H \cap Z$ defines the structure of a multiple point on $\{a\}$. If $\text{codim}_a Y = 2$, H can be considered as a 2-plane. Briançon [3] has classified all multiplicity structures on $0 \in \mathbb{C}^2$ up to multiplicity $\mu = 6$. We give the first cases of his list. For a suitable local coordinate system (x,y) at $0 \in \mathbb{C}^2$, thepossible idealsfor multiplicity ≤ 4 are

μ	I
1	(x,y)
2	(x,y^2)
3	(x,y^3), (x^2,xy,y^2)
4	(x,y^4), (x^2,y^2), (x^2,xy,y^3)

(0.4) A subspace Z of a manifold X is called a locally complete intersection, if for every point $a \in Z$ the ideal $I_{Z,a}$ can be generated by $r = \text{codim}_a Z$ elements. Locally complete intersections are Cohen-Macaulay.

In the sequel, we will often use the abbreviation CM for Cohen-Macaulay and l.c.i. for locally complete intersection.

§ 1. The Ferrand construction

In this section we recall the Ferrand construction [4] of the doubling of a l.c.i., since this is the basis for our later studies of higher multiplicities.

(1.1) Let $Y \subset X$ be a l.c.i. of codimension 2 in a manifold X. The sheaf $\nu_Y := I_Y/I_Y^2$ is then locally free of rank 2 over $\mathcal{O}_Y = \mathcal{O}_X/I_Y$, i.e. corresponds to a vector bundle of rank 2 on Y, which is by definition the conormal bundle of Y. (In the sequel we will identify vector bundles and locally free sheaves.) Now let there be given a line bundle L on Y, i.e. a locally free sheaf of rank 1, and an epimorphism

$$\beta : \nu_Y \longrightarrow L .$$

Then we can define an ideal $I_Z \subset \mathcal{O}_X$ with $I_Y^2 \subset I_Z \subset I_Y$ by the following exact sequence

$$(1) \qquad 0 \longrightarrow \frac{I_Z}{I_Y^2} \longrightarrow \nu_Y \longrightarrow L \longrightarrow 0 .$$

An easy calculation shows that I_Z is again locally generated by two elements: In a neighborhood of a point $y \in Y$ we may choose generators g_1, g_2 of $I_{Y,y}$ such that their classes $\dot{g}_i := g_i \bmod I_Y^2 \in \mathcal{V}_{Y,y}$ satisfy: $\beta(\dot{g}_1) = 0$ and $\beta(\dot{g}_2)$ is a generator of the stalk L_y. Therefore $(I_Z/I_Y^2)_y$ is generated by the class $\dot{g}_1$, hence

$$I_{Z,y} = (g_1) + I_{Y,y}^2 = (g_1, g_1^2, g_1g_2, g_2^2) = (g_1, g_2^2).$$

The subspace $Z = (|Y|, \mathcal{O}_X/I_Z)$ is called the Ferrand doubling of Y with respect to the epimorphism $\beta: \mathcal{V}_Y \longrightarrow L$. (The multiplicity of Z is twice the multiplicity of Y.)
It is clear that two epimorphisms $\beta: \mathcal{V}_Y \longrightarrow L$ and $\beta': \mathcal{V}_Y \longrightarrow L'$ define the same subspace Z iff there exists an isomorphism $\varphi: L \longrightarrow L'$ such that $\beta' = \varphi \circ \beta$.

(1.2) Since Z is again a l.c.i., the conormal sheaf $\mathcal{V}_Z = I_Z/I_Z^2$ is locally free, i.e. a vector bundle. We consider its restriction $\mathcal{V}_Z|Y := \mathcal{V}_Z \otimes \mathcal{O}_Y$. We have

$$\mathcal{V}_Z|Y = (I_Z/I_Z^2) \otimes (\mathcal{O}_X/I_Y) \cong I_Z/I_YI_Z .$$

On the other hand, by definition $L = I_Y/I_Z$, hence

$$L^2 = (I_Y/I_Z)^{\otimes 2} \cong I_Y^2/I_YI_Z .$$

Therefore we get an exact sequence

$$0 \longrightarrow L^2 \longrightarrow \mathcal{V}_Z|Y \longrightarrow I_Z/I_Y^2 \longrightarrow 0,$$

which can be fitted together with (1) to yield the following exact sequence of vector bundles on Y:

$$0 \longrightarrow L^2 \longrightarrow \mathcal{V}_Z|Y \longrightarrow \mathcal{V}_Y \longrightarrow L \longrightarrow 0.$$

From this it follows in particular that

(2) $\det(\mathcal{V}_Z|Y) = \det(\mathcal{V}_Y) \otimes L.$

This formula can be used to calculate the dualizing sheaf ω_Z of Z. The dualizing sheaf, which is just the canonical line bundle in the case of a manifold, can be calculated for a l.c.i. Z in a manifold X by the formula

$$\omega_Z = (\omega_X|Z) \otimes \det(\mathcal{V}_Z)^*.$$

Since a similar formula holds for ω_Y, we get from (2)

$$\omega_Z|Y = \omega_Y \otimes L^{-1}.$$

(1.3) If $Y \subset X$ is a submanifold and $Z \supset Y$ a CM-subspace with $|Z| = |Y|$ and multiplicity 2, one can conversely show that $I_Y^2 \subset I_Z \subset I_Y$ and $L := I_Y/I_Z$ is locally free of rank 1, hence Z is obtained from Y by the Ferrand construction by means of the natural epimorphism

$$\nu_Y = I_Y/I_Y^2 \longrightarrow I_Y/I_Z = L .$$

§ 2. Primitive extensions

(2.1) From now on, we consider always the following situation: Let Y be a smooth connected curve in a 3-dimensional manifold X. We are interested in Cohen-Macaulay subspaces Z of X with $Z \supset Y$ and $|Z| = |Y|$.

Such a CM subspace Z is called a <u>primitive extension</u> of Y if Z is locally contained in a smooth surface F.

Let us first study the local structure of a primitive extension. We may assume that there is a coordinate system (t,x,y) around the considered point such that F is given by $I_F = (x)$ and Y is given by $I_Y = (x,y)$. Since Z is a CM codimension 1 subspace of F, it is given in this coordinate system by $I_Z = (x,y^{k+1})$ for a certain natural number k. This shows that Z is even a l.c.i. (of multiplicity k+1).

To study the global structure of Z, we define a filtration

$$Y = Z_o \subset Z_1 \subset \dots \subset Z_k = Z$$

as follows: We denote by $Y^{(j)}$ the j-th infinitesimal neighborhood of Y in X, given by the ideal $I_{Y^{(j)}} = I_Y^{j+1}$ and set

$$Z_j := Z \cap Y^{(j)} , \quad \text{i.e.} \quad I_{Z_j} = I_Z + I_Y^{j+1}.$$

With respect to the local coordinates considered above, we have

$$I_{Z_j} = (x,y^{j+1}).$$

Thus Z_j is a l.c.i. of multiplicity j+1.

Let us assume $k \geqslant 1$. Then we have in particular the extension $Y \subset Z_1$ of multiplicity 2, which can be obtained by the Ferrand construction with the line bundle

$$L = I_Y/I_{Z_1} = I_Y/(I_Z + I_Y^2).$$

We will say in this situation that $Z \supset Y$ is a primitive extension of type L.

(2.2) Proposition. Let $Z \supset Y$ be a primitive extension of multiplicity $k+1$ and type L. Then one has for $j = 1,\dots,k$ exact sequences

$$0 \longrightarrow L^j \longrightarrow \mathcal{O}_{Z_j} \longrightarrow \mathcal{O}_{Z_{j-1}} \longrightarrow 0,$$

where $Z_j = Z \cap Y^{(j)}$. Further, with the abbreviation $I_j := I_{Z_j}$ one has isomorphisms

$$L^j \cong I_{j-1}/I_j \cong I_Y^j/I_1 I_Y^{j-1}.$$

Proof. We remark first that I_{j-1}/I_j is a locally free $\mathcal{O}_Y$-module of rank 1. This is verified by a local calculation. (In the above coordinates, I_{j-1}/I_j is generated by the class of y^j.) On the other hand, one has surjective $\mathcal{O}_Y$-morphisms

$$L^j = \left(\frac{I_Y}{I_1}\right)^{\otimes j} \xrightarrow{\varphi} \frac{I_Y^j}{I_1 I_Y^{j-1}} \xrightarrow{\psi} \frac{I_Z + I_Y^j}{I_Z + I_Y^{j+1}} = \frac{I_{j-1}}{I_j}$$

Since L^j and I_{j-1}/I_j are locally free of rank 1, φ and ψ have to be isomorphisms.

(2.3) Proposition. Let $Z \supset Y$ be a primitive extension of multiplicity $k+1$ and type L. Then there is an exact sequence

$$0 \longrightarrow L^{k+1} \xrightarrow{\tau} \nu_Z|Y \longrightarrow \nu_Y \longrightarrow L \longrightarrow 0.$$

The dualizing sheaf of Z satisfies

$$\omega_Z|Y = \omega_Y \otimes L^{-k}.$$

Proof. We have

$$\nu_Z|Y = (I_Z/I_Z^2) \otimes (\mathcal{O}_X/I_Y) \cong I_Z/I_Y I_Z$$

$$L = I_Y/I_1 = I_Y/(I_Z+I_Y^2),$$

$$L^{k+1} \cong I_Y^{k+1}/I_1 I_Y^k.$$

The inclusions

$$I_Y^{k+1} \subset I_Z \subset I_Y ,$$

$$I_1 I_Y^k \subset I_Y I_Z \subset I_Y^2 \subset I_1$$

induce the sequence we are looking for:

$$0 \longrightarrow \frac{I_Y^{k+1}}{I_1 I_Y^k} \longrightarrow \frac{I_Z}{I_Y I_Z} \longrightarrow \frac{I_Y}{I_Y^2} \longrightarrow \frac{I_Y}{I_1} \longrightarrow 0.$$

The exactness is again verified by local calculation. Taking determinants, we get from it

$$\det(\nu_Z|Y) = \det(\nu_Y) \otimes L^k.$$

This implies $\omega_Z|Y = \omega_Y \otimes L^{-k}$.

Remark. The above formula for ω_Z gives this line bundle only after restriction to Y. Thus one needes information about the restriction map Pic(Z) $\longrightarrow$ Pic(Y). For this we refer to § 3.2.

Now we study the following problem: Let there be given a primitive extension $Z' = Z_{k-1} \supset Y$ of multiplicity $k \geq 1$ and type L. Under what conditions can we extend further to a primitive extension $Z \supset Z' \supset Y$ of multiplicity k+1? Here we have

(2.4) Proposition. Let $Z' \supset Y$ be a primitive extension of type L and multiplicity k and let

$$\tau' : L^k \longrightarrow \nu_{Z'}|Y$$

be the natural injection (given by Proposition 2.3). Then there is a bijection between the set of primitive extensions $Z \supset Z' \supset Y$ of multiplicity k+1 and the set of retractions for τ', i.e. the set of epimorphisms

$$\beta : \nu_{Z'}|Y \longrightarrow L^k$$

with $\beta \circ \tau' = id_{L^k}$.

This correspondence is given by the sequence

$$(2) \qquad 0 \longrightarrow \frac{I_Z}{I_Y I_{Z'}} \xrightarrow{\alpha} \frac{I_{Z'}}{I_Y I_{Z'}} = \nu_{Z'}|Y \xrightarrow{\beta} L^k \longrightarrow 0.$$

<u>Proof.</u> a) Suppose first given a retraction β for τ' and define I_Z by the exact sequence (2). That $Z \supset Z' \supset Y$ is a primitive extension of multiplicity k+1 can be seen locally: In suitable coordinates,

$$I_Y = (x,y), \qquad I_{Z'} = (x,y^k).$$

In the considered neighborhood, a basis of the bundle $\nu_{Z'}|Y$ is constituted by the classes $\dot{x},\dot{y}^k$ of x,y^k modulo $I_Y I_{Z'}$, and $L^k = I_Y^k/I_1 I_Y^{k-1}$ is generated by $e := y^k \bmod I_1 I_Y^{k-1}$. Since β is a retraction, we have

$$\beta(\dot{y}^k) = e, \quad \beta(\dot{x}) = ce.$$

Replacing x by $x' = x - cy^k$, we have $I_Y = (x',y)$, $I_{Z'} = (x',y^k)$ and $\beta(\dot{x}') = 0$. Then Ker β is generated by the class of x', hence

$$I_Z = (x') + I_Y I_{Z'} = (x',y^{k+1}),$$

which shows that Z is a primitive extension of multiplicity k+1.

b) Conversely, if $Z \supset Z' \supset Y$ is a primitive extension of multiplicity k+1, we have $I_Z \subset I_Y I_{Z'}$ and

$$\mathrm{Im}\left(\frac{I_Z}{I_Y I_{Z'}} \xrightarrow{\alpha} \frac{I_{Z'}}{I_Y I_{Z'}}\right)$$

is a subline bundle of $\nu_{Z'}|Y$, which is the complement of the subline bundle $\mathrm{Im}(\tau') \subset \nu_{Z'}|Y$ (this is verified by a local calculation). Hence the epimorphism of $\nu_{Z'}|Y$ to the cokernel of α can be identified with the projection of $\nu_{Z'}|Y$ onto the summand $\mathrm{Im}(\tau') \cong L^k$ in the direct sum decomposition $\nu_{Z'}|Y = \mathrm{Im}(\alpha) \oplus \mathrm{Im}(\tau')$.

c) It is clear that different retractions $\beta_1,\beta_2 : \nu_{Z'}|Y \longrightarrow L^k$ define different ideals I_{Z_1}, I_{Z_2}.

Remark. For the sequence

$$0 \longrightarrow L^k \xrightarrow{\tau'} \nu_{Z'}|Y \longrightarrow \nu_Y \longrightarrow L \longrightarrow 0$$

let $M := \mathrm{Ker}(\nu_Y \longrightarrow L)$. This is a line bundle with $M = \det(\nu_Y) \otimes L^{-1}$. The existence of a retraction for τ' is equivalent to the splitting of the sequence

$$0 \longrightarrow L^k \longrightarrow \nu_{Z'}|Y \longrightarrow M \longrightarrow 0.$$

Therefore we obtain

(2.5) Corollary. Let $Z' \supset Y$ be a primitive extension of type L and multiplicity $k \geq 2$.

a) A sufficient condition for the existence of a primitive extension $Z \supset Z' \supset Y$ of multiplicity k+1 is

$$H^1(Y, \det(\nu_Y)^* \otimes L^{k+1}) = 0.$$

b) If there exists one primitive extension $Z^o \supset Z' \supset Y$ of multiplicity k+1, then the set of all primitive extensions $Z \supset Z' \supset Y$ of multiplicity k+1 is in bijective correspondence with

$$H^o(Y, \det(\nu_Y)^* \otimes L^{k+1}).$$

§3. Cohen-Macaulay filtrations, quasi-primitive extensions

(3.1) Let Y be a smooth connected curve in a 3-dimensional manifold X and $Z \supset Y$ a CM subspace of X with $|Z| = |Y|$. We will first define the Cohen-Macaulay filtration of the extension $Z \supset Y$. If $Y^{(j)}$ denotes the j-th infinitesimal neighborhood of Y, the intersection $Z \cap Y^{(j)}$ will not be necessarily Cohen-Macaulay, since in the primary decomposition of $I_{Z \cap Y^{(j)}}$ there might be embedded components. Throwing away all these embedded components, we get a well-defined largest CM subspace

$$Z_j \subset Z \cap Y^{(j)}.$$

Let $k \in \mathbb{N}$ be minimal with $Z \subset Y^{(k)}$, (since Y is connected, k exists). Then of course $Z = Z_k$. The sequence

$$Y = Z_o \subset Z_1 \subset Z_2 \subset \ldots \subset Z_k = Z$$

is called the CM-filtration of Z. One has always $I_Y^{j+1} \subset I_{Z_j}$ and there exists a 0-dimensional subset $S \subset Y$ such that

$$I_{Z_j,y} = I_{Z,y} + I_{Y,y}^{j+1} \quad \text{for all } y \in Y \setminus S \text{ and } j = 0,\dots,k.$$

For abbreviation let us write $I_j := I_{Z_j}$. We assert that

$$I_Y I_{j-1} \subset I_j .$$

This is trivially true in all points $y \in Y \setminus S$, hence $(I_Y I_{j-1} + I_j)/I_j$ is an ideal in $\mathcal{O}_{Z_j}$ with support contained in S. Since $\mathcal{O}_{Z_j}$ is CM, this ideal must be identically zero, which proves our assertion. Therefore

$$L_j := I_{j-1}/I_j$$

are modules over $\mathcal{O}_Y$, which are torsion-free (since $\mathcal{O}_{Z_j}$ is CM), hence locally free. Thus $Z = Z_k$ can be obtained from $Y = Z_o$ by successive extensions

$$(1) \qquad 0 \longrightarrow L_j \longrightarrow \mathcal{O}_{Z_j} \longrightarrow \mathcal{O}_{Z_{j-1}} \longrightarrow 0, \quad j = 1,\dots,k,$$

by vector bundles L_j. The multiplicity of Z is therefore

$$\mu(Z) = 1 + \sum_{j=1}^{k} \operatorname{rank}(L_j)$$

and we have

$$\chi(Z,\mathcal{O}_Z) = \chi(Y,\mathcal{O}_Y) + \sum_{j=1}^{k} \chi(Y,L_j).$$

<u>(3.2)</u> Since $L_j = I_{j-1}/I_j$ is an ideal of square zero in $\mathcal{O}_{Z_j}$, we get from (1) exact sequences

$$0 \longrightarrow L_j \longrightarrow \mathcal{O}_{Z_j}^{*} \longrightarrow \mathcal{O}_{Z_{j-1}}^{*} \longrightarrow 0,$$

hence exact sequences

$$H^1(Y,L_j) \longrightarrow \operatorname{Pic}(Z_j) \longrightarrow \operatorname{Pic}(Z_{j-1}) \longrightarrow H^2(Y,L_j)$$

from which one can read off sufficient cohomological conditions for the bijectivity of the restriction map $\operatorname{Pic}(Z) \longrightarrow \operatorname{Pic}(Y)$.

(3.3) Analogously to the formula $I_Y I_{j-1} \subset I_j$ one proves $I_i I_j \subset I_{i+j+1}$ for all i,j. This induces a natural multiplicative structure

$$L_i \otimes L_j \longrightarrow L_{i+j} .$$

In particular, one has morphisms

$$L_1^{\otimes j} \longrightarrow L_j ,$$

which are surjective over $Y \setminus S$.

(3.4) We have always a surjective map

$$\nu_Y = \frac{I_Y}{I_Y^2} \longrightarrow \frac{I_Y}{I_1} = L_1 .$$

Hence $\mathrm{rank}(L_1) \leqslant \mathrm{rank}(\nu_Y) = 2$. The case $\mathrm{rank}(L_1) = 0$ is trivial, since this implies $L_j = 0$ for all $j > 0$, hence $Z = Y$. So there remain two non-trivial cases:

i) $\mathrm{rank}(L_1) = 1$,
ii) $\mathrm{rank}(L_1) = 2$.

The second case occurs iff $I_1 = I_Y^2$, i.e. $Y^{(1)} \subset Z$.
In the first case we will call the extension $Z \supset Y$ quasi-primitive. Since generically (i.e. over $Y \setminus S$) we have $I_1 = I_Z + I_Y^2$, the condition $\mathrm{rank}(L_1) = 1$ is equivalent to the condition that generically $\mathrm{emdim}_y Z = 2$. Thus $Z \supset Y$ is a quasi-primitive extension iff it is a primitive extension outside a zero-dimensional subset of Y.

(3.5) Let now $Z \supset Y$ be a quasi-primitive extension with CM-filtration

$$Y = Z_o \subset Z_1 \subset \dots \subset Z_k = Z$$

and define the bundles $L_j = I_{j-1}/I_j$ as above. We will use the abbreviation $L := L_1$. Since the maps $L^j \longrightarrow L_j$ are generically surjective, it follows that all L_j are line bundles and that there are divisors $D_j \geqslant 0$ on Y such that

$$L_j = L^j(D_j).$$

From the multiplication $L_i \otimes L_j$ —— L_{i+j} we get

$$D_i + D_j \leqslant D_{i+j} \quad \text{for all } i,j \geqslant 1,$$

where $D_1 := 0$.
Thus to any quasi-primitive extension $Z \supset Y$ we can associate as invariants a line bundle L and a sequence of divisors $D_2,\dots,D_k$ on Y. We call $(L,D_2,\dots,D_k)$ the <u>type</u> of the quasi-primitive extension.

<u>(3.6)</u> Note that the extension $Z_1 \supset Y$ is obtained by the Ferrand construction using the line bundle L. The other extensions have a more complicated structure. To study them consider the conormal sheaves $\nu_j := \nu_{Z_j} = I_j/I_j^2$. We have $\nu_j|Y = I_j/I_YI_j$. Since $I_YI_j \subset I_{j+1}$ and $L_{j+1} = I_j/I_{j+1}$ we have an exact sequence

$$0 \longrightarrow \frac{I_{j+1}}{I_YI_j} \longrightarrow \frac{I_j}{I_YI_j} = \nu_j|Y \xrightarrow{\beta_j} L_{j+1} \longrightarrow 0.$$

Thus I_{j+1} is uniquely determined by I_j and the epimorphism $\beta_j\colon \nu_j|Y \longrightarrow L_{j+1}$. However this epimorphism is not arbitrary, but satisfies a certain condition. To derive this condition, we consider the sequence

$$0 \longrightarrow L^{j+1} \xrightarrow{\tau_j} \nu_j|Y \longrightarrow \nu_Y \longrightarrow L \longrightarrow 0.$$

As in § 2.3 we have $L^{j+1} = I_Y^{j+1}/I_1I_Y^j$, $\nu_j|Y = I_j/I_YI_j$, $\nu_Y = I_Y/I_Y^2$, $L = I_Y/I_1$ and the maps are induced by the natural inclusions. The sequence is a complex, but not necessarily exact at the places $\nu_j|Y$ and ν_Y. The composition

$$L^{j+1} \xrightarrow{\tau_j} \nu_j|Y \xrightarrow{\beta_j} L_{j+1} = L^{j+1}(D_{j+1})$$

is nothing else than the natural inclusion $L^{j+1} \longrightarrow L^{j+1}(D_{j+1})$. Thus β_j is a "meromorphic" retraction of τ_j. In a sense, this is the only condition that β_j has to fulfill, as the following proposition shows.

<u>(3.7) Proposition.</u> Let $Z' \supset Y$ be a quasi-primitive extension of type $(L,D_2,\dots,D_{k-1})$ and multiplicity k and let $\tau'\colon L^k \longrightarrow \nu_{Z'}|Y$ be the natural map induced by the inclusion $I_Y^k \subset I_{Z'}$. Let $D_k \geqslant 0$ be another divisor on Y. Then there exists a natural bijective correspondence between the set of quasi-primitive extensions $Z \supset Y$ of multiplicity $k+1$ and type $(L,D_2,\dots,D_k)$ with CM-filtration $Y = Z_0 \subset \dots \subset Z_{k-1}=Z' \subset Z$ and the set of all epimorphisms

$$\beta : \nu_{Z'}|Y \longrightarrow L^k(D_k)$$

which make commutative the diagram

$$\begin{array}{ccc} \nu_{Z'}|Y & \xrightarrow{\ \beta\ } & L^k(D_k) \\ \tau' \uparrow & \nearrow_{\text{nat.}} & \\ L^k & & \end{array}$$

Proof. Of course, given β, the associated extension $Z \supset Y$ is defined by the exact sequence

$$0 \longrightarrow \frac{I_Z}{I_Y I_{Z'}} \longrightarrow \nu_{Z'}|Y \xrightarrow{\ \beta\ } L^k(D_k) \longrightarrow 0.$$

By the above remarks it remains only to show that for this Z the maximal CM subspace of $Z \cap Y^{(k-1)}$ coincides with Z'. This is true over $Y \setminus \bigcup \operatorname{Supp}(D_j)$, since there the extension is primitive. Hence it is true everywhere.

(3.8) Parametrization. Assume Y compact. Then, given one β_o satisfying the conditions of Proposition 3.7, the set of all such β is in bijective coorespondence with an open subset of

$$\operatorname{Hom}(K, L^k(D_k)),$$

where $K := (\nu_{Z'}|Y)/\operatorname{Im}(L^k \longrightarrow \nu_{Z'}|Y)$. To determine this set consider the sequence

$$0 \longrightarrow L^k \xrightarrow{\ \tau'\ } \nu_{Z'}|Y \longrightarrow \nu_Y \longrightarrow L \longrightarrow 0.$$

Since this sequence is exact outside a set of dimension zero, $K' := K/\operatorname{Tors}(K)$ is isomorphic to

$$\operatorname{Im}(\nu_{Z'}|Y \longrightarrow \nu_Y) \subset M := \operatorname{Ker}(\nu_Y \longrightarrow L) = I_1/I_Y^2.$$

It follows that $K' = M(-D'_{k-1})$, where D'_{k-1} is the divisor determined by

$$\frac{I_1}{I_{k-1}+I_Y^2} \cong \mathcal{O}_{D'_{k-1}}.$$

Since $\operatorname{Hom}(K, L^k(D_k)) = \operatorname{Hom}(K', L^k(D_k))$ and $M = \det(\nu_Y) \otimes L^{-1}$, we see that the set of all β's is parametrized by an open subset of

$$H^o(Y, \det(\nu_Y)^* \otimes L^{k+1}(D'_{k-1}+D_k)),$$

(cf. Corollary 2.5).

Note that $D'_1 = 0$ and that

$$\mathcal{O}_{D_j} = \mathrm{Coker}(L^j \longrightarrow L_j) \cong \frac{I_{j-1}}{I_j + I_Y^j},$$

hence in particular $D'_2 = D_2$.

<u>(3.9) Local structure.</u> Proposition 3.7 can also be used to determine the local structure of quasi-primitive extensions. As an example consider a quasi-primitive extension $Z = Z_2 \supset Z_1 \supset Y$ of multiplicity 3 in the neighborhood of a point $a \in Y$ where $\mathrm{ord}_a(D_2) = d > 0$. Since $Z_1 \supset Y$ is a Ferrand doubling, there exists a local coordinate system (t, x, y) around a such that $I_Y = (x, y)$, $I_1 = (x, y^2)$ and $t(a) = 0$. Then $\nu_1|Y = I_1/I_Y I_1$ is generated by the classes

$$\dot{x} := x \bmod I_Y I_1, \qquad \dot{y}^2 := y^2 \bmod I_Y I_1,$$

and $L^2 = I_Y^2/I_Y I_1$ is generated by $\dot{y}^2$. In the diagram

$$\begin{array}{ccccccccc} 0 & \longrightarrow & \dfrac{I_2}{I_Y I_1} & \longrightarrow & \dfrac{I_1}{I_Y I_1} & \xrightarrow{\ \beta\ } & L^2(D_2) & \longrightarrow & 0 \\ & & & & \uparrow{\scriptstyle \tau} & \nearrow{\scriptstyle \gamma} & & & \\ & & & & L^2 & & & & \end{array}$$

τ maps $\dot{y}^2$ to $\dot{y}^2$ and γ maps $\dot{y}^2$ to $t^d e$, where e is a local base of $L^2(D_2)$. By the commutativity of the diagram $\beta(\dot{y}^2) = t^d e$. Since β is surjective, we must have $\beta(\dot{x}) = \varphi e$, where $\varphi(0) \neq 0$. Replacing x by $\frac{1}{\varphi}x$, we may suppose $\beta(\dot{x}) = e$. Then $\mathrm{Ker}(\beta)$ is generated by $t^d \dot{x} - \dot{y}^2$, hence

$$I_{Z_2} = (t^d x - y^2) + I_Y I_1 = (t^d x - y^2, xy, x^2).$$

In a similar manner one calculates the local structure of a quasi-primitive extension $Z = Z_3 \supset Z_2 \supset Z_1 \supset Y$ of multiplicity 4 and type (L, D_2, D_3) around a. One gets:

i) If $\mathrm{ord}_a(D_2) = \mathrm{ord}_a(D_3) = d$, then Z_3 is a l.c.i. in a neighborhood of a and

$$I_{Z_3} = (t^d x - y^2, x^2).$$

If globally $D_2 = D_3 =: D$, then Z_3 is a l.c.i. everywhere and one calculates for the dualizing sheaf

$$\omega_{Z_3}|Y = \omega_Y \otimes L^{-3}(-D).$$

ii) If $\mathrm{ord}_a(D_2) = d < \mathrm{ord}_a(D_3) = d + \delta$, then Z_3 is not a l.c.i. and in suitable coordinates

$$I_{Z_3} = (t^\delta(t^d x - y^2) - xy,\ y(t^d x - y^2),\ x^2).$$

§ 4. Thick extensions of multiplicity 4

(4.1) As always, let Y be a smooth connected curve in a 3-dimensional manifold X. A CM-extension $Z \supset Y$ which is not quasi-primitive contains by § 3.4 the full first infinitesimal neighborhood $Y^{(1)}$ of Y. Therefore we will call it a thick extension. In particular, if $Z \supset Y$ is a thick CM-extension of multiplicity 4, we have $Y^{(1)} \subset Z \subset Y^{(2)}$, i.e.

$$I_Y^3 \subset I_Z \subset I_Y^2$$

and $L := I_Y^2/I_Z$ is locally free of rank 1. Thus we have an exact sequence

$$0 \longrightarrow \frac{I_Z}{I_Y^3} \longrightarrow \frac{I_Y^2}{I_Y^3} \longrightarrow L \longrightarrow 0.$$

Conversely, let L be a given line bundle on Y and

$$\lambda: I_Y^2/I_Y^3 = S^2\nu_Y \longrightarrow L$$

an epimorphism. Then $\mathrm{Ker}(\lambda)$ can be written in the form I_Z/I_Y^3 and I_Z defines a CM-extension $Z \supset Y$ of multiplicity 4 with $Y^{(1)} \subset Z$.

(4.2) We study now the problem under what conditions on λ the structure Z will be l.c.i. For this purpose we consider more generally a bundle F of rank 2 on Y. One has the squaring map

$$q: F \longrightarrow S^2F.$$

Its image is a qudratic cone $Q \subset S^2F$. If e_1, e_2 is a local base of F and e_1^2, e_1e_2, e_2^2 the associated base of the second symmetric power S^2F, then Q consists of all linear combinations $\xi_1 e_1^2 + \xi_2 e_1 e_2 + \xi_3 e_2^2$ such that $4\xi_1\xi_3 - \xi_2^2 = 0$. Let now

$$\lambda : S^2F \longrightarrow L$$

be an epimorphism of S^2F onto a line bundle L on Y. We define a discriminant disc(λ) as follows: Let e be a basis of L over some open subset $U \subset Y$ and let e_1, e_2 be a basis of F over U as above. Then λ defines functions a,b,c on U by

$$\lambda(e_1^2) = ae, \quad \lambda(e_1e_2) = be, \quad \lambda(e_2^2) = ce.$$

With respect to the given bases, disc(λ) is given by $ac - b^2$. The transformation behavior under base changes of F and L shows then, that disc(λ) is a well defined element

$$\mathrm{disc}(\lambda) \in \Gamma(Y, \det(F)^{-2} \otimes L^2).$$

The discriminant has the following significance: disc(λ) vanishes in a point $p \in Y$ if and only if in the fiber S^2F_p the kernel $\mathrm{Ker}(\lambda)_p$ is tangent to the quadratic cone. Now we apply this to the bundle $F = \nu_Y$.

<u>(4.3) Proposition.</u> Let $\lambda : S^2\nu_Y \longrightarrow L$ be an epimorphism onto a line bundle L on Y and let $Z \supset Y$ be defined by the exact sequence

$$0 \longrightarrow \frac{I_Z}{I_Y^3} \longrightarrow S^2\nu_Y \xrightarrow{\lambda} L \longrightarrow 0.$$

Then Z is a l.c.i. at a point $p \in Y$ iff $\mathrm{disc}(\lambda)(p) \neq 0$.

<u>Proof.</u> a) If $\mathrm{disc}(\lambda)(p) \neq 0$, then in the fiber $(S^2\nu_Y)_p$ the kernel $\mathrm{Ker}(\lambda)_p$ intersects the quadratic cone in two different lines. Therefore there exist over some neighborhood of p two subline bundles $M_1, M_2 \subset \nu_Y$ such that $Q \cap \mathrm{Ker}(\lambda) = q(M_1) \cup q(M_2)$. Chose a basis e_1, e_2 of ν_Y such that e_i is a basis of M_i. Then Ker(λ) is generated by e_1^2 and e_2^2. We can choose local coordinates (t,x,y) in X around p such that $e_1 = x \bmod I_Y^2$ and $e_2 = y \bmod I_Y^2$. Then it is easily verified that $I_Z = (x^2, y^2)$, so Z is a l.c.i. in a neighbohood of p.

b) If $\operatorname{disc}(\lambda)(p) = 0$, we have to distinguish two cases:

i) $\operatorname{disc}(\lambda)$ vanishes identically in a neighborhood of p. This implies $Q \cap \operatorname{Ker}(\lambda) = q(M)$ for some subline bundle M of ν_Y over a neighborhood of p. Then for some basis $e_1 \in M$, e_2 of ν_Y, $\operatorname{Ker}(\lambda)$ is generated by e_1^2, e_1e_2. For a suitable coordinate system (t,x,y) around p we have then

$$I_Z = (x^2,xy) + (x,y)^3 = (x^2,xy,y^3),$$

which shows that Z is not a l.c.i.

ii) $\operatorname{disc}(\lambda)$ vanishes at p of a certain finite order $d > 0$. If (a,b,c) are the coordinates of λ with respect to some basis e_1,e_2 of ν_Y and e of L over a neighborhood of p, we have therefore $ac - b^2 = t^d$, where t is a local coordinate on Y with $t(p) = 0$. Since a,b,c cannot simultaneously vanish at p, we have $a(p) \neq 0$ or $c(p) \neq 0$. We may suppose $a(p) \neq 0$. Multiplying e_1 by an invertible function, we may even assume $a \equiv 1$. We replace now e_2 by $e_2' = e_2 - be_1$. Then

$$\lambda(e_1e_2') = \lambda(e_1e_2) - b\lambda(e_1^2) = b - b = 0.$$

Hence we may also assume without loss of generality that $b = 0$. Then $c = t^d$ and $\operatorname{Ker}(\lambda)$ is generated by e_1e_2, $e_2^2 - t^d e_1^2$. For a suitable coordinate system (t,x,y) around p we have then

$$I_Z = (xy,y^2-t^dx^2) + (x,y)^3 = (xy,y^2-t^dx^2,x^3),$$

which shows again that Z is not a l.c.i.

<u>(4.4) Remark.</u> From Proposition 4.3 it follws in particular: If Z is a locally complete intersection everywhere, then the bundle $\det(\nu_Y)^{-2} \otimes L^2$ must be trivial.

As an example let us consider the case $X = \mathbb{P}_3$, $Y = \mathbb{P}_1 \subset \mathbb{P}_3$. Then $\nu_Y = \mathcal{O}_Y(-1) \oplus \mathcal{O}_Y(-1)$. Thus for a thick l.c.i. structure $Z \supset Y$ of multiplicity 4 we have $L = \mathcal{O}_Y(-2)$. The epimorphism

$$S^2\nu_Y = \mathcal{O}_Y(-2)^3 \xrightarrow{\ \lambda\ } \mathcal{O}_Y(-2)$$

is then given by a triple of constants a,b,c with $ac - b^2 \neq 0$ and it is easy to see that there exist (global) homogeneous coordinates (u,v,x,y) on $\mathbb{P}_3$ such that

$$I_Y = (x,y), \quad I_Z = (x^2,y^2).$$

Thus Z is a global complete intersection.

(4.5) Proposition. Let $Z \supset Y$ be a thick l.c.i. etension of multiplicity 4 given by an epimorphism $\lambda : S^2\nu_Y \longrightarrow L$. Then we have for the dualizing bundle $\omega_Z|Y \cong \omega_Y \otimes L^{-1}$.

Proof. There is an epimorphism

$$\nu_Z|Y = I_Z/I_YI_Z \longrightarrow I_Z/I_Y^3 = \mathrm{Ker}(S^2\nu_Y \text{ --- } L),$$

which must be an isomorphism, since both sheaves are locally free $\mathcal{O}_Y$-modules of rank 2. Thus we have an exact sequence

$$0 \longrightarrow \nu_Z|Y \longrightarrow S^2\nu_Y \longrightarrow L \longrightarrow 0,$$

from which it follows that

$$\det(\nu_Z|Y) \cong \det(S^2\nu_Y) \otimes L^{-1} \cong \det(\nu_Y)^3 \otimes L^{-1}.$$

Since Z is a l.c.i., we have $\det(\nu_Y)^2 = L^2$, hence $\det(\nu_Z|Y) \cong \det(\nu_Y) \otimes L$, from which the assertion follows.

Bibliography

[1] C. Bănică and N. Manolache: Rank 2 stable vector bundles on $P^3(\mathbb{C})$ with Chern classes $c_1=1$, $c_2=4$. Math. Z. 190 (1985)315-339.

[2] C. Bănică and M. Putinar: On complex vector bundles on rational threefolds. Proc. Cambridge Math. Soc. 97 (1985)279-288.

[3] J. Briancon: Description of $\mathrm{Hilb}^n\mathbb{C}\{x,y\}$. Invent. Math. 41(1977) 45-89.

[4] D. Ferrand: Courbes gauches et fibrés de rang 2. C.R. Acad. Sci. Paris 281 (1977) 345-347.

[5] R. Hartshorne: Stable vector bundles of rank 2 on P^3. Math. Ann. 238 (1978) 229-280.

[6] L. Szpiro: Equations defining space curves. Tata Institute Lecture Notes. Bombay 1979.

C. Bănică
Department of Mathematics
INCREST
bd. Pacii 220
79622 Bucharest
Roumania

O. Forster
Mathematisches Institut
der LMU
Theresienstr. 37
D-8000 München 2
West Germany

Contemporary Mathematics
Volume 58, Part I, 1986

ALGEBRAIC CYCLES AND THE BEILINSON CONJECTURES

Spencer Bloch[1]

The purpose of this note is to explain a sort of generalization of the notion of algebraic cycle, and to relate these generalized cycles to higher K-theory and to various conjectures (Beilinson [2],[3], Beilinson-Lichtenbaum [4],[16] and Soulé [20]) connecting K-groups to L-functions and other objects of number-theoretic interest. I want to thank the organizers of the conference for providing a stimulating mathematical atmosphere.

The discussion here differs slightly from my talk at the conference for selfish reasons relating to my own publication record. Detailed verifications of the foundational properties of the groups $CH^*(X,n)$ will appear in Advances in Math. [26]. Since most of these properties are eminently believable and their proofs are modifications of classical arguments for cycles and for K-groups, I have simply listed them here with only the vaguest hints of proofs. On the other hand, the simple but powerful construction of cycle classes occurred to me more recently. Since it gives considerable information on the torsion theory, as well as a reformulation of Beilinson's conjectures relating K-groups to values of L-functions, I have chosen to treat it in more detail.

¶1. THE BASIC CONSTRUCTION. Let X be a scheme of finite type over a field k. We assume X is equidimensional, and we fix an integer $r \geq 0$. The cycle group $z^r(X)$ is defined to be the free abelian group with generators the irreducible closed subvarieties $Y \subset X$ of codimension r. We will say that subvarieties $Y_i \subset X$ of codimension r_i, $i = 1,2$ *meet properly* if every irreducible component of $Y_1 \cap Y_2$ has codimension $\geq r_1 + r_2$ on X.

Let $\Delta^n = \mathbf{A}^n_k$, $n = 0,1,2,\ldots$ with coordinates $(t_0,\ldots,t_n)$ satisfying $\Sigma t_i = 1$. For $I \subset \{0,1,\ldots,n\}$, a set of $n-m$ elements, we have a *face map*

$$\partial_I : X \times \Delta^m \hookrightarrow X \times \Delta^n$$

defined by setting $t_i = 0$, $i \in I$. We define $z^r(X,n)$ to be the free abelian group generated by irreducible codimension r subvarieties of $X \times \Delta^n$ meeting all faces $X \times \Delta^m$ for all $m < n$ properly.

1980 Mathematics Subject Classification. 14 C 35, 14 F 15

1 Supported by NSF

Thus

$$z^r(X,n)\subset z^r(X\times A^n).$$

With notation as above, we define

$$\partial_I : z^r(X,n) \to z^r(X,m)$$

to be the pullback map. Note that the faces are complete intersections and cycles are assumed to meet faces properly, so for $V\subset X \times \Delta^n$ irreducible of cod. r representing an element of $z^r(X,n)$ we have

$$\partial_I(V) = [V\cap\partial_I(X\times\Delta^m)].$$

Here [scheme] means associated cycle ([8], Chap.1).

There are also defined flat maps (degeneracies)

$$\delta_i : X\times\Delta^{n+1} \to X\times\Delta^n$$

$$\delta_i(x,t_o,\ldots,t_{n+1}) = (x,t_o,\ldots,t_{i-1},t_i+t_{i+1},t_{i+2},\ldots,t_{n+1})$$

and pullback along the δ_i (defined for all cycles since the δ_i are flat) defines maps δ_i: $z^r(X,n) \to z^r(X,n+1)$. The groups $z^r(X,n)$ $n = 0,1,2,\ldots$ together with the ∂_I and δ_i form a simplicial complex of abelian groups, denoted $z^r(X,.)$

$$\cdots \;\substack{\to\\\to\\\to\\\to}\; z^r(X,2) \;\substack{\to\\\to\\\to}\; z^r(X,1) \;\substack{\to\\\to}\; z^r(X,o)$$

The homotopy (or homology, the two coincide by [26]) of this complex is denoted by

$$H_n(z^r(X,.)) = \Pi_n(z^r(X,.)) \underset{\text{def}}{=} CH^r(X,n).$$

This is our basic object of study.

¶2. PROPERTIES. Here is a list of the properties of the groups $CH^r(X,n)$ established in [26].

(i) *Functoriality:* Covariant (with shift of codimension index r) for proper maps, and contravariant for flat maps. Contravariant for all maps when the target is smooth.

(ii) *Homotopy:* $CH^r(X,n) = CH^r(V(E),n)$ for $E \to X$ a vector bundle.

(iii) *Localization:* Let $Y\subset X$ be closed of pure codimension d. Then there is a long exact sequence

$$\cdots \to CH^*(X-Y,n+1) \to CH^{*-d}(Y,n) \to CH^*(X,n) \to CH^*(X-Y,n) \to \cdots \to CH^*(X-Y,1) \to$$

$$\to CH^{*-d}(Y,0) \to CH^*(X,0) \to CH^*(X-Y,0) \to 0.$$

(iv) *Degree zero:* $CH^r(X,0) = CH^r(X)$, the Chow group as defined in [8].

(v) *Local to global spectral sequence:* $CH^r(X,n) = \mathbb{H}^{-n}(X,\underline{\underline{z}}^r_X(.))$. where $\underline{\underline{z}}^r_X(.)$ is the complex of Zariski sheaves concentrated in negative degrees given by $U \to z^r(U,.)$. In particular, given $r \geq 0$ there is a spectral sequence $E_2^{p,q} = H^p(X,\underline{\underline{CH}}^r(-q)) \Rightarrow CH^r(X,-p-q)$

where $\underline{\underline{CH}}^r(q)$ is the Zariski sheaf associated to the presheaf $U \to CH^r(U,q)$.

(vi) *Multiplicativity:* There is an exterior product

$$CH^p(X,q) \otimes CH^r(Y,s) \to CH^{p+r}(X \times Y, q+s).$$

Pulling back along the diagonal yields a product structure

$$CH^p(X,q) \otimes CH^r(X,s) \to CH^{p+r}(X,q+s)$$

for X smooth.

(vii) *Chern classes:* For E on X a rank n vector bundle, there are well defined operators $\underline{\underline{c}}_i(E): CH^a(X,b) \to CH^{a+i}(X,b)$, $1 \leq i \leq n$ having the functoriality properties detailed in (Fulton [8] chap. 3). In particular, writing $\underline{\underline{\xi}}$ for the first Chern class of $\mathcal{O}(1)$ on $\mathbb{P}(E) \overset{\Pi}{\to} X$, one has the *projective bundle theorem*

$$\left(\bigoplus_{i=0}^{n-1} \underline{\underline{\xi}}^i \right) \circ \Pi^*: CH^*(X,m) \overset{\simeq}{\to} CH^*(\mathbb{P}(E),m)$$

as well as the usual Chern class identity

$$\underline{\underline{\xi}}^n + \underline{\underline{c}}_1 \underline{\underline{\xi}}^{n-1} + \cdots + \underline{\underline{c}}_n = 0.$$

We can define $c_i(E) = \underline{\underline{c}}_i(E)(X) \in CH^i(X,0)$. When X is non-singular, $\underline{\underline{c}}_i$ = multiplication by c_i.

(viii) *Relationship with algebraic K-theory:* Let $G_*(X)$ denote the algebraic K-groups defined using the category of coherent sheaves on X, [17]. Soulé [19] has defined a γ - filtration on the $G_*(X)$. We have

$$CH^p(X,q) \otimes \mathbb{Q} \simeq gr^p_\gamma \, G_q(X) \otimes \mathbb{Q}.$$

(ix) *Codimension 1:* For X regular,

$$CH^1(X,q) \cong \begin{cases} Pic(X) & q = 0 \\ \Gamma(X,\mathcal{O}_X^*) & q = 1 \\ 0 & q \geq 2. \end{cases}$$

(x) *Gersten's conjecture:* For X smooth over k there are flasque resolutions

$$0 \to \underline{\underline{CH}}^r_X(q) \to \bigoplus_{x \in X^0} i_x CH^r(Sp\, k(x),q) \to \bigoplus_{x \in X^1} i_x CH^{r-1}(Sp\, k(x),q-1) \to$$
$$\cdots \to \bigoplus_{x \in X^q} i_x CH^{r-q}(Sp\, k(x),0) \to 0.$$

Here X^q denotes the set of points x on the scheme X whose Zariski closure $\overline{\{x\}}$ has codimension q, $i_x A$ for an abelian group A denotes the constant sheaf with stalk A supported on $\overline{\{x\}}$, and $\underline{\underline{CH}}^r_X(q)$ is the Zariski sheaf as in (v). In particular,

$$CH^r(X) = H^r(X, \underline{\underline{CH}}^r(r)).$$

(xi) *Finite coefficients and the étale topology:* Let $z_{X,ét}(\cdot)$ be the complex of sheaves on X for the étale topology, given by $U \mapsto z^*(U,\cdot)$. Let n be an integer prime to the characteristic of k and let $\Pi: X \to Sp\, k$ be the structure map. Then the pullback

$$\Pi^*(z^*_{Sp\ k,\acute{e}t}(\cdot) \otimes \underline{Z}/n\underline{Z}) \to z^*_{X,\acute{e}t}(\cdot) \otimes \underline{Z}/n\underline{Z}$$

is a quasi-isomorphism.

¶3. COMMENTS ON PROOFS. The proofs of (i) - (v) use two elementary moving lemmas. A preliminary version of [26] contained some inaccuracies at this point, so I will give details here. (cf. [15],[25])

LEMMA. Let X be an algebraic k-scheme and G a connected algebraic k-group acting on X. Let $A,B \subset X$ be closed subsets. Assume the map $G\times A \to X$, $(g,a) \to g.a$ is dominant and all fibres have the same dimension. Then there exists an open non-empty $U \subset G$ such that for $g \in U$ the intersection $g(A) \cap B$ is proper.

PROOF. Consider the diagram

$$\begin{array}{ccccc} G & \longleftarrow & G\times A & \longrightarrow & X \\ & & \cup & \square & \cup \\ & & C & \longrightarrow & B \end{array}$$

where C is the indicated fibre product. Our hypothesis implies dim C = dim G + dim A + dim B - dim X. We may take for U the open set in G where the fibres of $C \to G$ have smallest dimension. Q.E.D.

LEMMA. Let hypotheses be as in the previous lemma. Assume moreover we are given an overfield $K \supset k$ and a K-morphism $\psi : X \to G$. Define $\phi: X \to X$ by $\phi(x) = \psi(x).x$. We suppose that ϕ is an isomorphism, and that there is a non-empty $V \subset X$ open and defined over k such that for any $x \in V$ a scheme point, we have for $\pi: X_K \to X_k$,

$$\text{tr. deg.}_k\ k(\psi(x),\pi(x)) \geq \dim.\ G.$$

Then $(A \cap V) \cap B$ is proper.

PROOF. We have the diagram

$$\begin{array}{ccccc} (V\cap A)_K & \xrightarrow{(\psi,\pi)} & G\times A_K & \longrightarrow & X \\ \cup & & \cup & & \cup \\ (V\cap A)_K \cap C & \longrightarrow & C & \longrightarrow & B \end{array}$$

As in the previous lemma, dim C = dim G + dim A + dim B - dim X. Also $(V\cap A)_K \cap C$ is identified under ϕ with $\phi(V\cap A) \cap B$. It therefore suffices to show the intersection $C \cap (V \cap A)$ is proper on GxA. We can also replace A by $V\cap A$ and ignore V.

We regard this as an intersection problem on GxA for C arbitrary. Replacing C by a hyperplane section, we may assume $C \cap A$ is zero dimensional. For the intersection to be improper we must have dim C < dim G. Let $z \in A$ be a scheme point such that $(\psi(a),\pi(a)) \in C$. We must have $\text{tr. deg.}_k\ k(\psi(a), \pi(a)) < \dim G$, contradicting the hypotheses. Q.E.D.

EXAMPLE 1. Suppose $X = X' \times A^1$, with G acting trivially on A^1. We are given $\psi_0: A^1_K \to G_K$ and an open $W_k \subset A^1_k$ such that for all scheme points $w \in W_K$, tr. $\deg._k(\psi_0(x)) \geq \dim G - 1$ and tr. $\deg._k\ \psi_0(w) = \dim G$ if w is algebraic over k. Then the map $\psi = \psi_0 \cdot \mathrm{pr}: X \to G$ satisfies the hypotheses of (1.2). Indeed the map $\phi: X = X'xA^1 \to X$ is $\phi(x,a) = (\psi_0(a)x,a)$ and is therefore an isomorphism.

As an application, we suppose given a connected algebraic k-group G acting on X and a morphism $\psi_0: A^1 \to G$ defined over K. Write $\psi(x,y) = (\psi_0(y)x,y)$ as above. Let $Y \subset X$ be locally closed and defined over k, and assume $GxY \to X$ is dominant with constant fibre dimension. Let $z^*_Y(X,n) \subset z^*(X,n)$ be generated by irreducible subvarieties meeting all faces $Yx\Delta^m$ properly. We define a homotopy $h.: z^*(X,.) \to z^*(X_K,.+1)$ by composing

$$z^*(X_k,.) \to z^*_K(X_K,.) \to z^*(XxA^1_K,.) \overset{\psi}{\to} z^*(XxA^1_K,.) \overset{T.}{\to} z^*(X_K,.+1)$$

Here T_n is induced by a "triangulation" $A^1 \times \Delta^n \cong \Delta^{n+1}$ coming from affine maps $\Delta^{n+1} \underset{\cong}{\overset{\to}{\to}} A^1 \times \Delta^n$. Although T_n of a cycle does not generally meet all faces properly, the composite h. is defined. Assuming $\psi_0(0) = \mathrm{id.} \in G$ and $\psi_0(x)$ is k-generic for all $x \in A^1(\bar{k})$, $x \neq 0$, we find that $h.(z^*_Y(X_k,.)) \subset z^*_Y(X_K,.+1)$ and $\psi(1)^*(z^*(X_k,.)) \subset z^*_Y(X_K,.)$. Moreover h. gives a homotopy between $\psi(0)^* = \mathrm{id.}$ and $\psi(1)^*$. When K is purely transcendental over k, a specialization argument enables us to eliminate K and deduce the inclusion $z^*_Y(X,.) \subset z^*(X,.)$ is a quasi-isomorphism.

EXAMPLE 2. Take $X = ZxA^1$, $Y = Zx\{0,1\}$, $G = G_a$ acting on A^1. We see that

$$z^*_{Zx\{0,1\}}(ZxA^1,.) \cong z^*(ZxA^1,.).$$

This is the key point in the proof of the homotopy theorem (ii).

EXAMPLE 3. Take $X = P^N$, $Y \subset P^N$ quasi-projective. Then

$$z^*_Y(P^N,.) \cong z^*(P^N,.).$$

Given $W \subset Y$ locally closed, one can define a complex $z^*_{Y,W}(P^N,.) \subset z^*_Y(P^N,.)$ fitting into an exact sequence

$$0 \to z^*_{Y,W}(P^N,.) \to z^*_Y(P^N,.) \oplus z^*_W(Y,.) \overset{a}{\to} z^*(Y,.).$$

One shows as above that the inclusions

$$z^*_{Y,W}(P^N,.) \subset z^*_Y(P^N,.) \subset z^*(P^N,.)$$

are quasi-isomorphisms, and that a_n is surjective for a range of n which goes to infinity with N. This is the key point in the proof of the localization sequence (iii) as well as the general contravariant functoriality (i).

The relation between the $CH^r(X,n)$ and K-theory explained in (viii) is established in several steps.

STEP 0. By a localization argument, one reduces to the case X smooth and affine.

STEP 1. Let $Y_1, Y_2, \ldots, Y_s \subset X$ be smooth Cartier divisors such that the intersection $Y_I = \bigcap_{i \in I} Y_i$ is transversal for any $I \subset \{1,2,..,s\}$. Let $Z \subset X$ be a closed subscheme meeting all the Y_I properly. We define relative K-groups with supports in Z

$$K_*^Z(X;Y_1,\ldots Y_s)$$

and cycle groups

$$CH^{*,Z}(X;Y_1,..,Y_s;m).$$

There is defined a chern character

$$ch: K_m^Z(X;Y_1,\ldots,Y_s) \to \bigoplus_i CH^{i,Z}(X;Y_1,\ldots,Y_s;m) \otimes \mathbb{Q}.$$

The group on the right is a module for $CH^*(X,0)$, so one can define the action of multiplication by the Todd class $Td(X)$, and a map

$$\tau = (Td(X)\cdot)\cdot ch$$

STEP 2. Using work of Kratzer [13], and Soulé [19], one defines Adams operations on the $K_*^Z(X;Y_1,..,Y_s)$. Assuming X equi-dimensional and Z of pure codimension d with irreducible components Z_i, one shows

$$gr^d K_0^Z(X;Y_1,..,Y_s) = \{\Sigma\, n_i Z_i = z \mid z\cdot Y_j = 0,\ 1 \le j \le s\} = \{\text{cod. d } \textit{relative cycles supp. on } Z\}.$$

Thus, by dropping supports, one gets a cycle class map

$$\{\text{cod. d rel. cycles supp. on } Z\} \to gr_\gamma^d\, K_0(X;Y_1,..,Y_s).$$

STEP 3. We apply the above with X replaced by $X\times\Delta^m$ and $Y_i = X\times\Delta_i^{m-1}$ the i-th codimension 1 face. The cycle map now factors through

$$CH^d(X\times\Delta^m;\ X\times\Delta_0^m,\ldots;0) \underset{\tau}{\overset{\text{cycle}}{\rightleftarrows}} gr_\gamma^d\, K_0(X\times\Delta^m;\ X\times\Delta_0^{m-1},\ldots) \otimes \mathbb{Q}$$

and $\tau \circ \text{cycle} = \text{id}$.

STEP 4. "cycle" is surjective. To see this, we rewrite

$$K_0(X\times\Delta^n;\ X\times\Delta_0^{n-1},\ldots) = K_0(X\times S^n)/K_0(X),$$

where S^n denotes the "algebraic sphere"

$$S^n = \Delta_+^n \coprod_{\cup \Delta^{n-1}} \Delta_-^n$$

One has $K_0(X\times S^n) = \varinjlim K_0(Y_a)$, the limit being taken over the category of morphisms $X\times S^n \to Y_a$ with Y_a smooth. Classes in $K_0(Y_a)_{\mathbb{Q}}$ can be interpreted in terms of cycles in the usual way, and these cycles moved to general position and then pulled back to $X\times S^n$. Such a pulled back cycle z has traces $z_\pm$ on $X\times\Delta_\pm^n$. Clearly, $z_+ - z_-$ is a relative cycle on $X\times\Delta^n$ and so represents a class in $CH^*(X\times\Delta^n;\ X\times\Delta_0^{n-1},\ldots;0)$. One checks that this cycle maps to the desired class in $K_0(X\times S^n)/K_0(X) \otimes \mathbb{Q}$.

¶4. THE CYCLE CLASS. The next step is to study cycle classes associated to elements in $CH^r(X,n)$. For simplicity, we work in the category of smoooth quasi-projective varieties over a field, and our cycle classes take values in suitable cohomology theories. The extension to arbitrary quasi-projective varieties over a field is easily done, using cohomology with supports, and yields cycle classes in homology theories, as one would expect.

The two "cohomology theories" we have in mind are the Deligne-Beilinson theory $H^a_D(X,b)$ and the étale theory $H^a_{\acute{e}t}(X,b) = H^a_{\acute{e}t}(X,Z_\ell(b))$. Note that these theories are bigraded and are finer than the usual Weil cohomologies. For example, in the étale theory, we do not extend scalars to the separable closure of the ground field as one does to define the étale Betti groups. The Deligne-Beilinson theory is defined in the first instance for varieties over $\mathbb{C}$, and fits in a long exact sequence

$$\cdots \to H^{a-1}_{DR}(X/\mathbb{C}) \to H^a_D(X,b) \to H^a(X,\mathbb{Z}(b)) \oplus F^bH^a(X,\mathbb{C}) \to H^a_{DR}(X/\mathbb{C}) \to \cdots$$

One can define the theory for varieties over $\mathbb{R}$ by taking invariants (in a suitable derived sense) under the real conjugation.

We assume given such a bigraded theory $H^a(X,b)$ satisfying certain reasonable properties (listed below), and we define a cycle map

$$CH^b(X,n) \to H^{2b-n}(X,b).$$

The integer b will be fixed throughout the discussion. We assume $H^*(X,b)$ is hypercohomology (for the Zariski topology, the étale topology, or some other Grothendieck topology having enough points so Godement style resolutions are available) of a complex $K_X(b)^\cdot$ which is contravariant functorial in the sense that given $f:Y \to X$, there is a well-defined map $f^*K_X(b) \to K_Y(b)$ satisfying the usual compatibilities for more elaborate diagrams. Replacing $K_X(b)$ by its Godement resolution, we may assume the sheaves $K_X(b)^n$ are acyclic, so

$$H^a(X,b) = H^a(\Gamma(K_X(b)^\cdot)).$$

Because the Hodge filtration on an open variety is usually defined using the complex of logarithmic differentials on the compactification, it is not obvious that the Deligne-Beilinson cohomology can be calculated from such a complex. For a proof that it can, cf. [2], 1.6.

Given Y a (not necessarily smooth) closed subscheme of X, we assume there are defined cohomology groups with supports, and a localization sequence

$$\cdots \to H^a_Y(X,b) \to H^a(X,b) \to H^a(X-Y,b) \to H^{a+1}_Y(X,b) \to \cdots$$

which is contravariant for cartesian squares

$$\begin{array}{ccc} Y' & \hookrightarrow & X' \\ \downarrow & & \downarrow \\ Y & \hookrightarrow & X \end{array}$$

and covariant for $Y_1 \subset Y_2 \subset X$. Note we do not assume $K_{X-Y}(b) = K_X(b)|_{X-Y}$ but only that there is a map. Concerning these groups, we suppose
(homotopy): $H^*(X,b) = H^*(X \times A^1, b)$.

(cycle): Assume $Y \subset X$ has pure codimension b. Then there is a well defined cycle class

$$[Y] \in H^{2b}_Y(X,b).$$

This class is contravariant functorial in the sense that given $f:X' \to X$ such that $f^{-1}(Y)$ is pure codimension b in X', we have $f^*[Y] = [f^*Y]$ (pullback of cycles).
(weak purity): If $Y \subset X$ has pure codimension r, then $H^a_Y(X,b) = (0)$ for $a < 2r$.

To construct the cycle classes, we associate to our diagram

$$X \rightrightarrows X \times \Delta^1 \overset{\to}{\underset{\to}{\to}} X \times \Delta^2 \cdots$$

a spectral sequence (from the double complex $\Gamma(K_{X \times \Delta^{-p}}(b)^q)$

$$E_1^{p,q} = H^q(X \times \Delta^{-p},b) \Rightarrow ?.$$

Note there is a convergence problem, but using the homotopy axiom we find

$$E_2^{p,q} = \begin{cases} H^q(X,b) & p = 0 \\ 0 & p \neq 0. \end{cases}$$

Since the d_1 boundary maps are either 0 or isomorphisms, we get the same result if we truncate at $X \times \Delta^N$ (i.e. terminate our diagram at $\vdots\ X \times \Delta^N$) for N even. Thus taking N large and even we may write

$$E_1^{p,q} \Rightarrow H^*(X,b).$$

Let $\dot{H}^a(X \times \Delta^p, b) = \varinjlim H^a_{|Z|}(X \times \Delta^p, b)$, where Z runs through $z^b(X,p)$ and $|Z|$ = Supp (Z). Assuming we have truncated at some large even N to avoid convergence problems, we get another spectral sequence

$$.E_1^{p,q} = \dot{H}^q(X \times \Delta^{-p},b), (-p \leq N.)$$

and there is a map of spectral sequences $.E_1^{p,q} \to E_1^{p,q}$. Also, the cycle class gives us a map of complexes (t_N means truncate in degree N)

$$t_N z^b(X,.) \to (.E_1^{-.,2b}, d_1).$$

Note for $r > 1$, $d_r : .E_r^{p,2b} \to .E_r^{p+r,2b-r+1} = 0$ (weak purity), so we get $CH^b(X,n) \to .E_2^{-n,2b} \to .E_\infty^{-n,2b}$. Again by weak purity, $.E_\infty^{x,y} = 0$ for $y < 2b$, so we get

$$CH^b(X,n) \to .E_\infty^{-n,2b} \to \{\text{limit of } .E_1 \text{ spectral sequence in deg. } 2b-n\}$$

$$\to \{\text{limit of } E_1 \text{ spectral sequence in deg. } 2b-n\} = H^{2b-n}(X,b).$$

This is the desired cycle map. Further conditions on the cohomology theory (satisfied in cases of interest) guarantee compatibility with the product structure.
EXAMPLE. We define

$$CH^r(X,\mathbb{Z}/N\mathbb{Z};n) = H_n(z^r(X,.)\otimes \mathbb{Z}/N\mathbb{Z})$$

An easy variant on the above construction gives cycle maps

$$CH^r(X,Z/NZ;n) \to H^{2r-n}_{\acute{e}t}(X,Z/NZ(r))$$

for N prime to the characteristic. Using the codimensional 1 structure from ¶1. (ix) we get

$$\Gamma(X,\mu_N)^{\otimes r} = CH^1(X,Z/NZ;2)^{\otimes r} \to CH^r(X,Z/NZ;2r) \to \Gamma(X,\mu_N^{\otimes r}).$$

The composition is the natural map so for example if the N-th roots of unity are k, we see

$$CH^r(X,Z/NZ;2r) = \mu_N^{\otimes r} \oplus ?$$

¶5. BEILINSON CONJECTURES. Consider now the cycle class in Deligne-Beilinson cohomology. More precisely, we suppose X smooth and projective over a number field k, and we consider map (*regulator map*).

$$CH^r(X_k,n) \to CH^r(X \underset{\mathbb{Q}}{\times} \mathbb{C},n) \to H_D^{2r-n}(X \underset{\mathbb{Q}}{\times} \mathbb{C},r).^+$$

(The + means invariant under real conjugation.)

For example if $X = \mathrm{Sp}\,k$ and $r = n = 1$ we get the standard map

$$k^* \to (k\otimes\mathbb{C})^{*+} = (k\otimes\mathbb{R})^* \underset{\log|\ |}{\to} \mathbb{R}^{r_1+r_2}$$

This example suggests some relation with Zeta and L-functions, as in the classical regulator of a number field. We would like for the image of the regulator map to be discrete, so we must replace k^* by $\mathcal{O}_k^*$, the units, and more generally X by X, a global regular model (assumed to exist) for X over $\mathrm{Sp}\mathcal{O}_k$. Note that the groups $CH^r(X,n)$ can be defined (although in the absence of appropriate moving lemmas, not much can be said about them) and we can hope the regulator maps

$$CH^r(X,n) \overset{R}{\to} H_D^{2r-n}(X \underset{\mathbb{Q}}{\times} \mathbb{C},r)^+$$

have discrete images. If we factor out by the maximal compact subgroup on the right, we get

$$\bar{R}:CH^r(X,n) \longrightarrow H^{2r-n-1}(X \underset{\mathbb{Q}}{\times} \mathbb{C}, \mathbb{R}(r-1))^+/F^rH_{DR}^{2r-n-1}(X \underset{\mathbb{Q}}{\times} \mathbb{R}/\mathbb{R})$$

$$\underset{\text{def.}}{=} H_D^{2r-n}(X \underset{\mathbb{Q}}{\times} \mathbb{R}, \mathbb{R}(r)).$$

Beilinson notes that the $\mathbb{R}$- vector space on the right has a natural "$\mathbb{Q}$-volume", i.e. a $\mathbb{Q}$-structure on $\Lambda^{\max}$ coming from the evident Q-structures on $H^{2r-n-1}(X \times \mathbb{C},\mathbb{R}(r-1))$ and $F^rH_{DR}^{2r-n-1}(X \underset{\mathbb{Q}}{\times} \mathbb{R}/\mathbb{R})$. He observes that the rank of

$H_D^{2r-n}(X \times \mathbb{C}, \mathbb{R}(r))$ equals the conjectural order of zero of the L-function $L(H^{2-r-1}_{\mathbb{Q}},s)$ associated to the $\mathrm{Gal}(\bar{k}/k)$-representation on $H^{2r-n-1}_{\text{ét}}(X_{\bar{k}},\mathbb{Q}_1)$, at the point s=r. Define $B^r(X)$ = cod. r cycles on X defined over k modulo homological equivalence, and write

$$M = \begin{cases} CH^r(X,n) & n \geq 2 \\ CH^r(X,1) \oplus B^{r-1}(X), & n=1 \end{cases}$$

Note when n=1, the target of $\bar{R}$ is $H^{r-1,r-1}(X \times \mathbb{R}, \mathbb{R}(r-1)$ so in all cases there is a map $\bar{R}: M \to H_D^{2r-n}(X \underset{\mathbb{Q}}{\times} \mathbb{R}, \mathbb{R}(r))$.

CONJECTURE (Beilinson). $\bar{R}$ embeds M, modulo torsion, as a lattice of maximal rank in $H_D^{2r-n}(X \times \mathbb{R}, \mathbb{R}(r))$. Moreover

$$R(\overset{\max}{\wedge} M \otimes \mathbb{Q}) = \ell \cdot \{\mathbb{Q}\text{-volume structure}\}$$

where ℓ is the first non-vanishing term in the Taylor expansion for $L(H^{2r-n-1},s)$ at s=r-n.

¶6. LICHTENBAUM CONJECTURES. Lichtenbaum [16] has conjectured the existence of complexes $\Gamma(r)$, r = 0,1,2,... in the derived category of étale sheaves on a scheme X. These complexes should satisfy a number of properties:

(0) $\Gamma(0) = \mathbb{Z}.$, $\Gamma(1) = \mathbb{G}_m[-1]$.

(1) For $r \geq 1$, $\Gamma(r)$ is acyclic outside of $[1,r]$.

(2) (Hilbert theorem 90) Let a_* be the functor which assigns to every étale sheaf of X the associated Zariski sheaf. Then the Zariski sheaf $R^{q+1}a_*\Gamma(q) = 0$.

(3) Let n be a positive integer prime to all residue field characteristics of X. Then there exists a triangle in the derived category of the form

$$\begin{array}{ccc} & Z/nZ(r) & \\ \swarrow & & \nwarrow \\ \Gamma(r) & \longrightarrow & \Gamma(r), \end{array}$$

where $Z/nZ(r)$ denotes the r-fold Tate twist of Z/nZ.

(4) There are product mappings $\Gamma(r) \times \Gamma(s) \to \Gamma(r+s)$, which induce maps on cohomology:

$$H^p(X,\Gamma(r)) \otimes H^q(X,\Gamma(s)) \to H^{p+q}(X, \Gamma(r+s)).$$

(5) The cohomology sheaves $H^i(\Gamma(r))$ are isomorphic to the étale sheafification of the functors $X \mapsto gr^r_\gamma K_{2r-i}(X)$ upto torsion involving primes less than or equal to r -1. Here gr_γ is the gradation corresponding to the Kratzer, Soulé

γ-filtration on higher K-theory. This isomorphism should come from an Atiyah-Hirzebruch spectral sequence which degenerates upto torsion involving primes $\leq r-1$.

(6) For F a field, $H^r(F,\Gamma(r))$ is canonically isomorphic to the Milnor K-groups $K_r^M(F)$.

Write $z^r_{\text{ét}}(X,.)$ for the complex of cycles sheafified for the étale topology, and define

$$\Gamma(r) = z^r_{\text{ét}}(X,2r-.).$$

CONJECTURE. Assume X is regular. Then the complex $\Gamma(r)$ satisfies (0)-(6) above.

At the moment, (0) is known, as is (4). The first part of (5) is true after tensoring with $\mathbb{Q}$, but I have no Atiyah-Hirzebruch spectral sequence. (3) is true upto a possible exotic direct factor, as sketched above. Conjectures (1) and (2) seem difficult, but (6) is probably more accessible.

Let me isolate a more elementary question which remains open for the $\Gamma(r)$ defined as above.

CONJECTURE. For $f:X \to Y$ proper of relative dimension n, there exists a trace morphism

$$tr : Rf_*\Gamma(n+m) \to \Gamma(m)[-2n]$$

satisfying the usual properties of a trace.

It is straightforward to define a mapping

$$f_*\Gamma(n+m) \to \Gamma(m)[-2n].$$

The problem, as O. Gabber pointed out to me, is to show this map factors through Rf_*. One can show that

$$Rf_*\Gamma(m) \otimes Q = f_*\Gamma(m) \otimes \mathbb{Q},$$

so tensoring $\Gamma(m)$ with the sequence $0 \to \mathbb{Z} \to \mathbb{Q} \to \mathbb{Q}/\mathbb{Z} \to 0$ it would suffice to exhibit a trace for the torsion complexes $\Gamma(n) \otimes \mathbb{Q}/\mathbb{Z}$.

PROPOSITION A. Assume the complexes $\Gamma(m)$ have traces. Let $p : \mathbb{P}^1 \times X \to X$ be projection. Then

$$Rp_*\Gamma_{\mathbb{P}^1 \times X}(r) = \Gamma_X(r) \oplus \Gamma_X(r-1)[-2].$$

PROPOSITION B. Let X be smooth over k and assume (n,char.k) = 1. Then, if a trace exists, we have

$$\Gamma(r)\otimes\mathbb{Z}/n\mathbb{Z} \cong \mu_n^{\otimes r}$$

PROOF of A. Let i: $X \to \mathbb{P}^1 \times X$ be the inclusion at ∞. We have

$$\Gamma_X(r-1)[-2] \xrightarrow{i_*} Rp_*\Gamma_{\mathbb{P}^1\times X}(r) \xrightarrow{p_*} \Gamma_X(r-1)[-2]$$

$$\Gamma_X(r) \xrightarrow{p^*} Rp_*\Gamma_{\mathbb{P}^1\times X}(r) \xrightarrow{i^*} \Gamma_X(r)$$

and $p_* i_* = \text{id.} = i^* p^*$. It follows that

$$Rp_*\Gamma_{\mathbb{P}^1\times X}(r) \cong \Gamma_X(r) \oplus \Gamma_X(r-1)[-2] \oplus ?,$$

the projections being give by p_* and i^*. To show ?=0, consider the diagonal $\Delta \subset P^1 \times P^1$. We have

$$[\Delta] = [\infty \times P^1] + [P^1 \times \infty] \in H^2(\mathbf{P}^1 \times \mathbf{P}^1, \Gamma(1)).$$

From the diagram

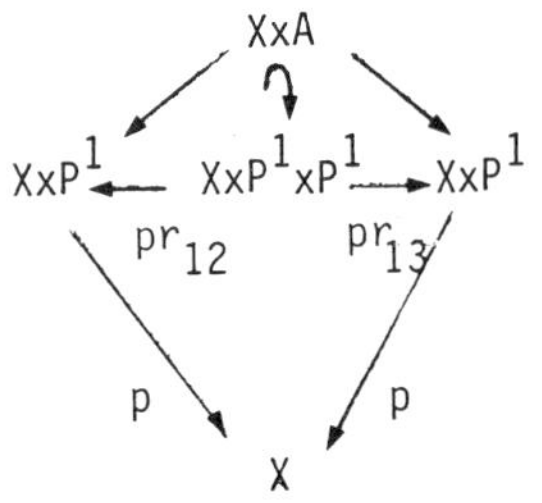

we compute the composite

$$Rp_*\,\Gamma_{X\times\mathbb{P}^1}(r) \xrightarrow[pr_{12}^*]{} Rpr_{1*}\Gamma_{X\times\mathbb{P}^1\times\mathbb{P}^1}(r) \xrightarrow[\cdot[A]]{} Rpr_{1*}\Gamma_{X\times\mathbb{P}^1\times\mathbb{P}^1}(r+1)[2]$$

$$\xrightarrow[pr_{13*}]{} Rp_*\Gamma_{X\times\mathbb{P}^1}(r)$$

with $A = \Delta$, $\infty \times P^1$, and $P^1 \times \infty$, respectively. For $A = \Delta$, we get the identity. For $A = \infty \times P^1$, we get $p^* i^*$, and for $A = P^1_{\times\infty}$ we find $i_* p_*$. Thus $\text{id} = p^* i^* + i_* p_*$, proving the proposition. Q.E.D.

PROOF of B. Let $p: \mathbf{P}^1 \times X \to X$. By Prop. A we have

$$(\Gamma_X(r)\otimes \mathbb{Z}/n\mathbb{Z}) \oplus (\Gamma_X(r-1) \otimes \mathbb{Z}/n\mathbb{Z}[-2]) \cong Rp_*(\Gamma_{\mathbb{P}^1 \times X}(r)) \overset{L}{\otimes} \mathbb{Z}/n\mathbb{Z}$$

$$\cong Rp_*(\Gamma_{\mathbb{P}^1 \times X}(r)\otimes \mathbb{Z}/n\mathbb{Z}) \cong Rp_*(p^*(\Gamma_X(r)\otimes\mathbb{Z}/n\mathbb{Z}))$$

$$\cong Rp_*(\mathbb{Z}/n\mathbb{Z}_{\mathbb{P}^1 \times X}) \underset{\mathbb{Z}/n\mathbb{Z}}{\overset{L}{\otimes}} (\Gamma_X(r)\otimes \mathbb{Z}/n\mathbb{Z})$$

$$\cong [(\mathbb{Z}/n\mathbb{Z})_X \oplus \mu_{n,X}^{-1} [-2]] \otimes (\Gamma_X(r) \otimes \mathbb{Z}/n\mathbb{Z}).$$

It follows that

$$\Gamma_X(r) \overset{L}{\otimes} \mathbb{Z}/n\mathbb{Z} \cong \Gamma_X(r-1) \overset{L}{\otimes} \mathbb{Z}/n\mathbb{Z} \underset{\mathbb{Z}/n\mathbb{Z}}{\otimes} \mu_n.$$

Since $\Gamma(0) = \mathbb{Z}$, we conclude by induction on n. Q.E.D.

REMARK. Assuming the complexes $\Gamma(m)$ have traces, one could use Prop. A. to define $\Gamma(m)$ for $m < 0$.

¶7. SOULE'S CONJECTURE. Finally and very briefly, I want to mention a conjecture of C. Soulé [20]. Let X be a scheme of finite type over $\mathbb{Z}$. Let $\zeta_X(s)$ denote the Zeta function of X, i.e.

$$\zeta_X(s) = \prod_x (1-N_x^{-s})^{-1}$$

where the product is taken over all closed points of X.

CONJECTURE. Suppose dim X = d. Then $\zeta_X(s)$ is meromorphic on $\mathbb{C}$, and

$$-\mathrm{ord}_{s=d-r}\zeta_X(s) = \sum_i (-1)^i \; \mathrm{rk}[CH^r(X,i)].$$

Of course, we do not know whether $CH^r(X,i)$ has finite rank, or whether $CH^r(X,i) \otimes \mathbb{Q} = 0$ for $i >> 0$. Soulé originally formulated the conjecture with $gr^r_\gamma K_i(X)$ in place of $CH^r(X,i)$. These two groups have the same rank by ¶2 (viii). The interesting point is that the right side looks like an Euler-Poincaré characteristic for the complex $z^r(X,.)$. Is there lurking here some sort of arithmetic index theory?

BIBLIOGRAPHY

[1] Baum, P., Fulton, W., MacPherson, R., Riemann-Roch for singular varieties, Publ.Math., I.H.E.S. 45 (1975),101-145.

[2] Beilinson, A., Higher regulators and values of L-functions (in Russian), Modern problems in mathematics VINIT series vol. 24 (1984),181-238.

[3] Beilinson, A., Height pairing between algebraic cycles, preprint (1984).

[4] Beilinson, A., Letter to C. Soulé, November 1, 1982.

[5] Bloch, S., Lectures on algebraic cycles, Duke University Math. series IV, 1980.

[6] Bloch, S., Algebraic K-theory and Zeta functions of elliptic curves, Proc. ICM, Helsinki,(1978), 511-515.

[7] Chevalley, C., Anneaux de Chow et applications, Sem. C. Chevalley 2é année, Secr.Math.Paris, (1958).

[8] Fulton, W., Intersection Theory, Springer Verlag Ergebnisse series (1984).

[9] Gabber, O., Preprint.

[10] Gersten, S., Some exact sequences in the higher K-theory of rings, in Algebraic K-theory I, Springer lecture notes 341, Springer-Verlag, (1973).

[11] Gillet, H., Riemann-Roch theorems for higher K-theory, Adv. in Math. 40 (1981), 203-289.

[12] Grothendieck et al. SGA VI, Springer lecture notes 225, Springer-Verlag, (1971).

[13] Kratzer, C., λ-structure en K-théorie algébrique, Comm.Math.Helv. No. 55, (1980),233-254.

[14] Landsburg, S., Relative cycles and algebraic K-theory, prerpint (1983).

[15] Levine, M., Cycles on singular varieties, preprint, (1983).

[16] Lichtenbaum, S., Values of Zeta functions at non-negative integers, preprint (1983).

[17] Quillen, D., Higher algebraic K-theory I, lecture notes in Math. no. 341, (1973), 85-147.

[18] Roberts, J., Chow's moving lemma, Appendix to exposé of S. Kleiman, Algebraic Geometry, Oslo, 1970, Wolters-Noordhoff publ Co. Groningen, (1972).

[19] Soulé, C., Operations en K-théorie Algébrique, to appear, J. Canadian Math.

[20] Soulé, C., K-théorie et zéros aux points entiers de fonctions zéta, Proc. ICM., Warszawa, (1983).

[21] Suslin, A. A., On the K-theory of algebraically closed fields, preprint.

[22] Suslin, A. A., On the K-theory of local fields, preprint.

[23] Dayton, B., and Weibel, C., A spectral sequence for the K-theory of affine glued schemes, Springer lectures notes in Math. 854 (1980), 24-92.

[24] Guillet, H., and Thomanson, The K-theory of strictly Hensel rings and a Theorem of Suslin, preprint.

[25] Kleiman, S., The transversality of a general translate, Comp. Math. 38 (1974), 287-297.

[26] Bloch, S., Algebraic cycles and higher K-theory, Advances in Math. (to appear).

Institut des Hautes Etudes Scientifiques
91440 Bures-sur-Yvette
France

Department of Mathematics
University of Chicago
Chicago, Illinois 60637
USA

Contemporary Mathematics
Volume 58, Part I, 1986

THE INFINITESIMAL ABEL-JACOBI MAPPING FOR HYPERSURFACES

Herbert Clemens

ABSTRACT. The derivative of the Abel-Jacobi mapping for a smooth hypersurface in $\mathbb{P}^4$ is computed in terms of the obstruction to splitting the normal bundle sequence for the triple (curve, threefold, $\mathbb{P}^4$). Applications to the cubic case are given.

1. ABEL-JACOBI MAPPING. Let

$$V \subseteq CP^4$$

be a smooth hypersurface of degree n. If

$$\mathcal{C} \longrightarrow S \times V$$

is a versal family of curves on V, with fibre C over some base point $s_0 \in S$, then the Zariski tangent space to S at s_0 can be identified with

$$H^0(C;N_{C,V}) ,$$

at least when $i: C \longrightarrow V$ is immersive, C non-singular, and

$$N_{C,V} = i^*T_V/T_C.$$

The Abel-Jacobi mapping associated to $\mathcal{C}$ is the morphism

$$\begin{aligned} \Phi: S &\longrightarrow J(V) = (H^{3,0}(V) + H^{2,1}(V))^*/H_3(V,\mathbf{Z}) \\ s &\longmapsto \int_C^{C_s} \quad , \end{aligned}$$

where

$$\int_C^{C_s} = \int_\Gamma \quad , \quad \partial\Gamma = C_s - C.$$

The derivative of Φ at s_0 is therefore given by a pairing

$$H^0(C; N_{C,V}) \otimes (H^0(\Omega^3{}_V) + H^1(\Omega^2{}_V)) \longrightarrow C.$$

Up to scalar, this pairing is just the obvious one induced by the contraction

$$N_{C,V} \otimes \Omega^i{}_V \longrightarrow \Omega^{i-1}_C \quad .$$

So, for example,

$$\Phi^*(H^{3,0}(V)) = 0.$$

The derivative of Φ at s_0 is what we mean by the *infinitesimal Abel-Jacobi mapping*.

2. THE CASE OF HYPERSURFACES. Following Griffiths [2] we denote by $\hat{\Omega}_{\mathbb{P}^4}(mV)$ the sheaf of closed r-forms on $(\mathbb{P}^4 - V)$ with no worse than m-th order poles along V, and we use these sheaves to produce 3-forms on V. Consider, now, the following commutative diagram

$$\begin{array}{ccccccccc}
0 & \longrightarrow & \hat{\Omega}^3_{\mathbb{P}^4}(V) & \longrightarrow & \Omega^3_{\mathbb{P}^4} & \xrightarrow{d} & \Omega^4_{\mathbb{P}^4}(2V) & \longrightarrow & 0 \\
 & & \downarrow & & \downarrow & & \downarrow & & \\
0 & \longrightarrow & \dfrac{\hat{\Omega}^3_{\mathbb{P}^4}(V)}{\hat{\Omega}^3_{\mathbb{P}^4}} & \longrightarrow & \dfrac{\Omega^3_{\mathbb{P}^4}(V)}{\Omega^3_{\mathbb{P}^4}} & \xrightarrow{\wedge \frac{df}{f}} & \dfrac{\Omega^4_{\mathbb{P}^4}(2V)}{\Omega^4_{\mathbb{P}^4}(V)} & \longrightarrow & 0 \\
 & \text{contraction} & \uparrow\cong & \text{contraction} & \uparrow\cong & & \uparrow N_{V,\mathbb{P}^4}\cong \mathcal{O}_V(V) & & \\
0 & \longrightarrow & T_V \otimes \Omega^4_{\mathbb{P}^4}(V) & \longrightarrow & T_{\mathbb{P}^4}|_V \otimes \Omega^4_{\mathbb{P}^4}(V) & \longrightarrow & N_{V,\mathbb{P}^4} \otimes \Omega^4_{\mathbb{P}^4}(V) & \longrightarrow & 0 \\
 & & \downarrow & \text{Residue} & \downarrow & \text{Residue} & \downarrow & & \\
0 & \longrightarrow & T_V \otimes \Omega^3_V & \longrightarrow & T_{\mathbb{P}^4}|_V \otimes \Omega^3_V & \longrightarrow & N_{V,\mathbb{P}^4} \otimes \Omega^3_V & \longrightarrow & 0 \\
 & & \downarrow & & \downarrow & & \downarrow & & \\
0 & \longrightarrow & N_{C,V} \otimes \Omega^3_V & \longrightarrow & N_{C,\mathbb{P}^4} \otimes \Omega^3_V & \longrightarrow & N_{V,\mathbb{P}^4}|_C \otimes \Omega^3_V & \longrightarrow & 0
\end{array}$$

where $f = 0$ is a local defining equation for V. This gives a commutative diagram

$$\begin{array}{ccc}
H^0(\Omega^4_{\mathbb{P}^4}(2V)) & \xrightarrow{\tau_V} & H^1(\hat{\Omega}^3_{\mathbb{P}^4}(V)) \\
\downarrow & & \downarrow \\
H^0(N_{V,\mathbb{P}^4}|_C \otimes \Omega^3_V) & \xrightarrow{\tau_C} & H^1(N_{C,V} \otimes \Omega^3)
\end{array}$$

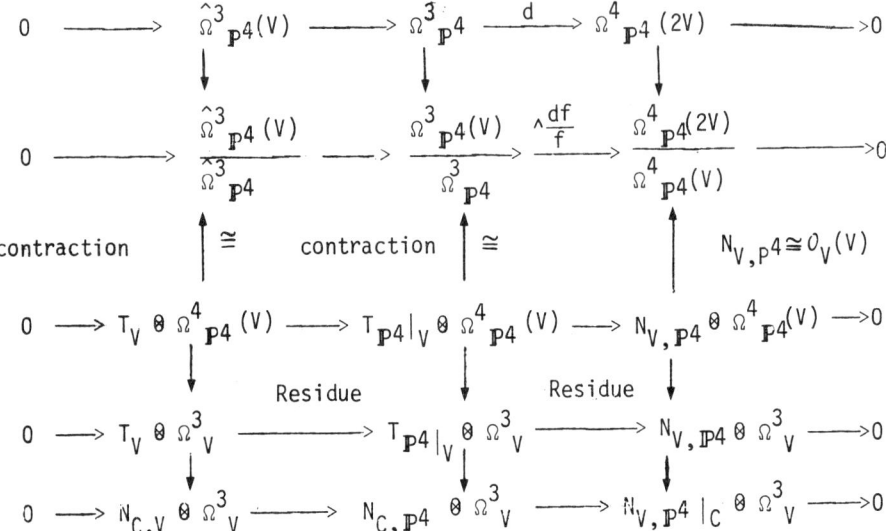

Now α is just the natural restriction map

$$\Omega^4_{\mathbb{P}^4}(2V) = \mathcal{O}_{\mathbb{P}^4}(2n-5) \longrightarrow N_{V,\mathbb{P}^4}|_C \otimes \Omega^3_V = \mathcal{O}_{\mathbb{P}^4}(2n-5)|_C.$$

Since contraction with tangent vectors pointing along V commutes with residue, the mapping β fits into a commutative diagram

$$\begin{array}{ccc} H^1(\hat{\Omega}^3_{\mathbb{P}^4}(V)) & \longrightarrow & H^1(\Omega^2_V) \\ \downarrow \beta & & \downarrow \Phi^* \\ H^1(N_{C,V} \otimes \Omega^3_V) & \underset{\substack{\text{Serre} \\ \text{duality}}}{=} & H^0(N_{C,V})^* \end{array}$$

This says that the infinitesimal Abel-Jacobi mappping is "given" by the mapping τ_C. This last mapping is just cup-product with the extension class

$$\{\sigma_{ij}\} \in \mathrm{Ext}^1(N_{V,\mathbb{P}^4}|_C, N_{C,V}) = H^1(N^*_{V,\mathbb{P}^4}|_C \otimes N_{C,V})$$

for the exact sequence

$$0 \longrightarrow N_{C,V} \longrightarrow N_{C,\mathbb{P}^4} \longrightarrow N_{V,\mathbb{P}^4}|_C \longrightarrow 0.$$

This computation was first given by Welters in [3, pp. 28-29].

3. COMPUTING THE OBSTRUCTION. We can make everything even more explicit as follows. Elements of $H^1(\Omega^2_V)$ are given by

$$\mathrm{Res}\,\frac{A\Omega}{F^2}, \quad A \in H^0(\mathcal{O}_{\mathbb{P}^4}(2n-5)),$$

where V is given (globally) by $F = 0$ and

$$\Omega = \left\langle \sum_{i=0}^{4} X_i \frac{\partial}{\partial X_i}, dX_0 \dots dX_4 \right\rangle .$$

Then, if $\nu \in H^0(N_{C,V})$,

$$\{A\nu \wedge \sigma_{ij}\}$$

is a Cech one-cochain on C with coefficients in

$$\mathcal{O}_C(n-5) \otimes \Lambda^2 N_{C,V} = \Omega^3_V|_C \otimes \Lambda^2 N_{C,V} \underset{\text{contraction}}{=} \Omega^1_C.$$

Thus

$$\Phi_* \left(\nu, \operatorname{Res} \frac{A\Omega}{F^2} \right) = \{A\nu \wedge \sigma_{ij}\}.$$

In special situations, the right-hand side of this formula is easy to compute. For example, suppose the image of C in $\mathbb{P}^4$ happens to be in a surface X so that

$$N_{C,V} \longrightarrow N_{X,\mathbb{P}^4}|_C$$

is an isomorphism off a finite set $\Sigma \subseteq C$. (X must be smooth at points of C.) If

$$p \in \Sigma ,$$

then X is tangent to V at p, so that its tangent space is given there by

$$dh_p = df = 0 ,$$

where $f = 0$ is a local defining equation for V. Let

$$U_p$$

be a small neighborhood of p in $\mathbb{P}^4$ with local coordinates

$$x_p , y_p , h_p , k_p$$

where $k_p = h_p = 0$ gives X locally,

$$y_p = h_p = f = 0$$

gives C locally, and

$$df|_p = dk_p|_p .$$

So there is a vector field ξ_p on U_p such that

$$\xi_p\Big|_p = \frac{\partial}{\partial k_p}\Big|_p \quad \text{and on } U_p , \ \langle \xi_p , df \rangle \equiv 1.$$

We can also write

$$f = y_p \ell_p(x,y) + h_p(\ldots) + k_p(\ldots)$$

This means that, in a punctured neighborhood of p in C,

$$\left\langle \frac{1}{\ell_p} \frac{\partial}{\partial y_p} , df \right\rangle \equiv 1 .$$

Next cover C by the following list of open sets

i) V_p such that $\overline{V}_p \subseteq U_p$, $p \in \Sigma$;

ii) $U'_p = (U_p \setminus \{p\})$, $p \in \Sigma$;

iii) $U \in \mathcal{U}$ such that $U \cap V_p = \emptyset$, $p \in \Sigma$.

The obstruction cochain $\{\sigma_{ij}\}$ in the formula for the infinitesimal Abel-Jacobi map can then be evaluated for this cover

$$\sigma_{U'_p} = \frac{1}{\ell_p} \frac{\partial}{\partial y_p} - \xi_p$$

with all other $\sigma_{U,U'} = 0$.

This says that if we pick a fixed section

$$B \in H^0(\mathbb{P}^4 ; \mathcal{O}_{\mathbf{P}^4}(2n-5))$$

such that B is never zero on any U_p , then , for each p , there is an invertible function

$$g_p \text{ on } U_p ,$$

depending only on the choice of B and the local coordinates on U_p and the choice of local defining equation f for V , such that

$$\Phi_*(\nu, \operatorname{Res} \frac{A\Omega}{F^2}) =$$

$$\sum_{p\in\Sigma} \operatorname{Res}_p g_p \cdot \frac{A}{B} < \nu \wedge (\frac{1}{\ell_p} \frac{1}{\partial y_p} - \xi_p) , dx_p dy_p dh_p > =$$

$$- \sum_{p\in\Sigma} \operatorname{Res}_p \frac{gp}{\ell_p} \cdot \frac{A}{B} <\nu , dh_p> dx_p$$

When C is the normalization of

$$C_0 = X \cdot V$$

where C_0 has only ordinary nodes, the above formula is more striking. In this case we may choose our local coordiantes around a node q to be such that near q

$$f|_X = x_q y_q$$

Let q' and q" be the preimages of q in C. Then the above formula becomes

$$\Phi_*(\nu, \frac{\operatorname{Res} A\Omega}{F^2}) = 2\pi i \sum_q \underbrace{g_q(q)}_{= C_q \neq 0} \frac{A(q)}{B(q)} <\nu_{q'} - \nu_{q''}, dh_q>.$$

4. THE CUBIC THREEFOLD. Consider the case in which $n = 3$, that is, V is a cubic hypersurface, and

$$\mathfrak{C} \longrightarrow S \times V$$

is the family of rational cubic curves on V. Notice that a curve C_s is either a smooth rational cubic spanning a $\mathbf{P}^3$ or C_s maps to a plane section of V. In this case,

$$H^{3,0}(V) = 0$$
$$H^{2,1}(V) = H^0(\mathbf{P}^4 ; (1))$$
$$\mathrm{Res}\,(A\Omega/F^2) \longleftarrow\!\!\dashv A$$

If C_s embeds in V, then it spans a hyperplane section of V a cubic surface, in which C_s moves in a two-dimensional linear system. Every hyperplane section of V contains such linear system. Thus:

S is an irreducible projective variety of dimension 6.
Also every immersive C_s which lies in a smooth hyperplane section of V has normal bundle sitting in a sequence

$$0 \longrightarrow \mathcal{O}_C(1) \longrightarrow N_{C,V} \longrightarrow \mathcal{O}_C(3) \longrightarrow 0$$

where $C_s \simeq C \times \mathbb{P}^1$ and the integers refer to degrees on $\mathbb{P}^1$. Thus

$$N_{C,V} \cong \begin{cases} \mathcal{O}_C(1) + \mathcal{O}_C(3) \\ \qquad \text{or} \\ \mathcal{O}_C(2) + \mathcal{O}_C(2) \end{cases}$$

Since $\dim H^0(N_{C,V}) = 6$ in both cases, the family S is versal and smooth at s. The same thing is true, by a similar analysis, if an immersive C_s lies in a hyperplane section of V with one ordinary node.

Let K be a plane section of V. We define the Abel-Jacobi mapping

$$\Phi : S \longrightarrow J(V)$$
$$s \longmapsto \int_K^{C_s}$$

(Since all plane sections K form a rational family, the choice of a particular K does not matter in definition of Φ.)

If $C = C_s$ is smooth and the hyperplane section Y of V spanned by C is also smooth, then C is linearly equivalent to a sum of three lines on Y:

$$L_1 + L_2 + L_3$$

where L_1 and L_3 span the hyperplane and L_2 meets L_1 and L_3. Let M be the line on Y such that

$$M + L_2 + L_3$$

give a plane section of Y. So

$$C - (\text{plane section}) = L_1 - M .$$

We can use this equivalence to compute the rank of Φ at s :

$$\Phi \text{ has fibre dimension } \geq 2 \text{ at } C_s$$

since $C = C_s$ moves in a two-dimensional linear equivalence class in Y. So

$$\text{rank}_s \quad \Phi \leq 4 .$$

To see that we have equality, it suffices to work with the lines L_1 and M given above. Let X be the union of two planes which contain $L_1 \cup M$ but which do not intersect at points of $L_1 \cup M$. X determines four points

$$q_1, q_2 \in L_1 \quad q_3, q_4 \in M$$

where the residual components of $(X \cdot V)$ meet L_1 and M. Then by the formula in §3,

$$\Phi_*\left(\nu , \frac{\text{Res } A\Omega}{F^2} \right) = \sum_{i=1}^{4} c_i \frac{A(q_i)}{B(q_i)} < \nu_{q_i} , dh_i>$$

where h_i defines X locally at q_i . Since X and ν can be chosen so that, for all i ,

$$< \nu_{q_i} , dh_i > \neq 0$$

and since $q_1,\ldots,q_4$ span a hyperplane in $\mathbb{P}^4$, the kernel of the codifferential of Φ is generated by the linear form A vanishing on $q_1,\ldots q_4$ and so on the hyperplane they span. Thus

$$\text{rank}_s \quad \Phi = 4 .$$

In fact, $\Phi(S)$ turns out to be the theta-divisor of the principally polarized abelian variety $J(V)$. To see this, specialize V to a cubic V_0 with a node. Projection from the node maps V_0 birationally onto $\mathbb{P}^3$ and S_0 onto the family of twisted cubics in $\mathbb{P}^3$ meeting a smooth canonical curve

$$D = \text{quadric} \cap \text{cubic} \subseteq \mathbb{P}^3$$

sic times. Since there is a unique twisted cubic through 6 generic points on a generic D, we have a commutative diagram

$$\begin{array}{ccc} S_0 & \xrightarrow{\quad \Phi_0 \quad} & J(V_0) \\ \updownarrow \ \text{birat.} & & \downarrow \lambda \\ D^{(6)} & \xrightarrow{\quad \Psi \quad} & J(D) \end{array}$$

(See [1] for further information about this construction.)

Since the fibres of Ψ are two dimensional and irreducible, the generic fibre of Φ_0 must consist of the cubics of a single linear equivalence class of a single hyperplane section of V_0. So the same must be true at a general fibre of

$$\Phi : X \longrightarrow J(V)$$

for a general cubic V. Also, the map λ in the above diagram must map $\Phi_0(S_0)$ birationally to $J(D)$ so that $\Phi_0(S_0)$ must be the theta divisor of the generalized abelian variety $J(V_0)$. This in turn, with some elementary observations about local properties of the mappings involved as V varies, leads to the conclusion, for general cubic V,

$$\Phi(S) = \text{theta divisor of } J(V).$$

Finally, we want to compute the tangent cone to $\Phi(S)$ at

$$0 \in J(V).$$

It is to this point that all rational plane cubics in V go under the Abel-Jacobi mapping Φ. The applicable formula is the one at the end of §3.

$$\Phi_*\left(\nu, \frac{\text{Res } A\Omega}{F^2}\right) = C\,\frac{A(q)}{B(q)} < \nu_{q'} - \nu_{q''}, dh_q > ,$$

where q is the node of an immersed rational plane cubic. Thus, if ν is a normal vector field which separates the local components of C at q, then

$$\text{Res } \frac{A\Omega}{F^2} \in H^{2,1}(V)$$

pairs to zero with ν if and only if

$$A(q) = 0 .$$

But that says that under the identification

$$\mathbb{P}^4 \simeq \mathbb{P}(H^{2,1}(V)^*)$$

$$A \longleftrightarrow \operatorname{Res} \frac{A\Omega}{F^2}$$

the point of q in $\mathbb{P}^4$ corresponds to the point

$$\Phi_*(\nu) \in H^{2,1}(V)^* .$$

Since, for any $q \in V$, the generic plane section of V by a plane which is tangent to V at q gives a curve C with many

$$\nu \in H^0(\tilde{C}; N_{\tilde{C},V})$$

of the above type, we conclude:

V is (a component of) the tangent cone to $\Phi(S)$ at $J(V)$.

(The possibility of other components of the tangent cone for generic V can be eliminated.) So classical facts about the singular locus of the theta divisor of the Jacobian of a curve imply that $J(V)$ is not a Jacobian of a curve, and so, by the criterion of [1], V is not rational.

BIBLIOGRAPHY

[1] Clemens, C. Herbert, and Griffiths, Phillip A., "The intermediate Jacobian of the cubic threefold, "Ann. Math., 95(1972), 281-356.

[2] Griffiths, Phillip A., "On the periods of certain rational integrals, I and II," Ann. Math., 90(1969), 460-541.

[3] Welters, G. E., Abel-Jacobi Isogenies for Certain Types of Fano Threefolds, Math. Centre Tracts 141, Mathematische Centrum, Amsterdam, 1981.

Department of Mathematics
University of Utah
Salt Lake City, Utah 84112
U.S.A.

Contemporary Mathematics
Volume 58, Part I, 1986

INFINITE COXETER GROUPS AND AUTOMORPHISMS OF ALGEBRAIC SURFACES

I.Dolgachev[1]

ABSTRACT. The group of automorphisms of an algebraic surface is represented naturally in the orthogonal group of the lattice of algebraic cycles. We study certain infinite Coxeter subgroups of the orthogonal group which play role in the description of the image of the representation of the automorphism group.

INTRODUCTION. The occurence of finite Coxeter groups in the theory of algebraic surfaces is notorious . The most striking example of this is the existence of 27 lines on a nonsingular projective cubic surface. The incidence preserving permutations of the set of these lines form a group isomorphic to the Weyl group $W(E_6)$ of a root system of type E_6. In their fundamental work [PS], I.Piatetsky-Shapiro and I.Shafarevich showed the importance of infinite Coxeter groups for the description of the automorphism group of a surface of type K3. In this talk we discuss their result in a slightly more general context and also present some new results on the computation of the automorphism group of Enriques and rational surfaces.

1. HOMOLOGY REPRESENTATIONS OF AUTOMORPHISMS. Let X be a nonsingular projective complex algebraic surface. The group Aut(X) of complex analytic (= regular algebraic) automorphisms of X is a topological group whose connected component of the identity $Aut(X)^0$ is a complex Lie group . Its dimension is equal to the dimension of the space of holomorphic vector fields on X. Let

$$A(X) = Aut(X)/Aut(X)^0 .$$

We are interested in the structure of this discrete group. In particular,we want to find when this group is finite and when it is infinite. As we will see , the first case occurs very often (in contrast to the case of algebraic curves, where it never occurs). Note that

1980 Mathematics Subject Classification 14J, 20F

[1] Research partially supported by a NSF grant

$$A(X) = \mathrm{Aut}(X)$$

except in the following cases:

a) X is an elliptic surface with smooth reduced fibres;

b) X is an abelian surface;

c) X is a ruled surface.

To study A(X) we consider its natural representation on 2-cycles on X. Let

$$H_X = H_2(X,\mathbf{Z})/\mathrm{torsion}$$

be the group of 2-dimensional homology of X modulo torsion equipped with the structure of a lattice (= bilinear symmetric form on a free abelian group) with respect to the intersection form. We denote by O(L) the orthogonal group (= the group of isometries) of a lattice L. Since $\mathrm{Aut}(X)^0$ acts identically on H_X, we have a natural representation

$$A(X) \to O(H_X), \quad g \mapsto g_*,$$

of A(X) in the orthogonal group of H_X.

By restriction, it defines a representation

$$r: A(X) \to O(S_X), \quad g \mapsto g_*,$$

in the sublattice S_X of H_X spanned by algebraic cycles. Recall that a 2-cycle is called algebraic if it is homologous to a 2-cycle of the form $\Sigma m_i[C_i]$, where $m_i \in \mathbf{Z}$ and $[C_i]$ is the homology class of an irreducible algebraic curve on X. Note that by the Hodge Index Theorem, S_X is a nondegenerate hyperbolic lattice ,i.e. the signature (t_+,t_-,t_0) of the corresponding real quadratic form is $(1,\rho-1,0)$, where $\rho = \rho(X) = \mathrm{rk}(S_X)$ is the Picard number of X.

Note also that there is a natural isomorphism between S_X and the Neron-Severi lattice

$$\mathrm{Num}(X) = \mathrm{Pic}(X)/\mathrm{Pic}^{\tau}(X),$$

where Pic(X) is the Picard group of isomorphism classes of divisors (or invertible sheaves, or complex line bundles on X) and $\mathrm{Pic}^{\tau}(X)$ its subgroup of numerically trivial classes.

There is a very distinguished vector in S_X which is preserved under g_* for any $g \in A(X)$. This is the image c_X of the first Chern class $c_1(X) \in H_2(X,\mathbf{Z})$ of X in S_X. Let

$$S_X' = (\mathbf{Z}c_X)^{\perp} \subset S_X,$$

the orthogonal complement of c_X in S_X. This is an even lattice of signature equal to

$$(1,\rho-2,0) \text{ if } c_X^2 < 0 \text{ or } c_X = 0,$$
$$(0,\rho-2,1) \text{ if } c_X^2 = 0\,,\ c_X \neq 0,$$
$$(0,\rho-1,0) \text{ if } c_X^2 > 0\,.$$

Let

$$r'\colon A(X) \to O(S_X')$$

be the restriction of r to S_X'. This is the main object of our study.

First, the following result is true for the kernel of r':

PROPOSITION 1. Ker(r') = Ker(r) is a finite group.

PROOF. Clearly Ker(r) = Ker(r'). Assume first that X is a regular surface (i.e. $q(X) = \dim H^1(X,\mathcal{O}_X) = 0$). Then the kernel of the canonical map $Pic(X) \to NS(X) \cong S_X$ is finite. Let H be a very ample divisor on X. Replacing H by nH, where n >> 0, we may assume that

$$g(H) \sim H \quad , \forall\, g \in Ker(r)',$$

where Ker(r)' is the preimage of Ker(r) in Aut(X). This implies that Ker(r)' is contained in the group of projective automorphisms of X, where the projective embedding is defined by the complete linear system |H|. The latter group is a projective linear group and, as such, has only finitely many connected components. This implies that Ker(r) is finite.

Assume now that q(X) > 0. Then, we consider different cases corresponding to the classification of algebraic surfaces.

Case 1: X is of general type.

Then, the assertion is obvious. As is well-known, the group Aut(X) is finite in this case.

Case 2. X is of Kodaira dimension 1.

Then, X has a unique structure of an elliptic surface $f\colon X \to B$ (see [BPV]). The group Aut(X) is naturally mapped to Aut(B) with the kernel Aut(X;f) isomorphic to the subgroup of automorphisms of the general fibre X_η over the field of functions $K = \mathbb{C}(B)$. Since Aut(B) has only finitely many connected components, it suffices to show that the kernel of the natural map $Aut(X;f) \to O(S_X)$ has only finitely many connected components. Clearly, Aut(X,f) is a subgroup of Aut(X',f'), where $f'\colon X' \to B'$ is an elliptic surface obtained from f by a base change $B' \to B$. The latter can be chosen in such a way that f' has a section, i.e. $X'_{\eta'}$ has a structure of an abelian variety of dimension 1. The group Aut(X',f') consists of translations and a finite group of automorphisms preserving a fixed zero section. Let $\mathrm{Aut}(X',f')_t$ be the subgroup of translations. For every section C of f', we have $C^2 = -\chi(X',\mathcal{O}_{X'}) \leq 0$ and the equality takes place if and only if f' is smooth (this easily follows from the standard facts on elliptic fibrations, see [BPV],Ch.V).

Assume f' is not smooth. This implies that the group of sections is mapped injectively into $S_{X'}$ and $\mathrm{Aut}(X';f')_t$ acts by translations on $S_{X'}$ modulo the sublattice spanned by components of fibres. Hence, any non-trivial $g \in \mathrm{Ker}(\mathrm{Aut}(X;f) \to O(S_X))$ cannot belong to $\mathrm{Aut}(X';f')_t$. Thus, the kernel is finite.

Assume that f' is smooth. Then, after a suitable base change, we may assume that $X' \cong E \times B'$, where E is an elliptic curve and B' is a curve of positive genus ([AS],Ch.VII,Thm.6). Since A(X') is finite, A(X) is finite in this case.

Case 3: X is an abelian surface.

Let L be a very ample line bundle on X. For every $g \in \mathrm{Ker}(\mathrm{Aut}(X) \to O(S_X))$,

$$g^*(L)\otimes L^{-1} \in \mathrm{Pic}_0(X),$$

where $\mathrm{Pic}_0(X)$ is the subgroup of isomorphism classes of line bundles algebraically equivalent to 0. It follows from [M],Ch.II,§8,Thm.1, that

$$g^*(L)\otimes L^{-1} = t_a{}^*(L)\otimes L^{-1},$$

where t_a is a translation automorphism of X corresponding to a point $a \in X$.

Thus, $(g \cdot t_{-a})^*(L) \cong L$, that shows that g belongs to a subgroup of Aut(X) which is generated by translations and automorphisms which leave the isomorphism class of L invariant. This is an algebraic subgroup of Aut(X) which has finitely many connected components.

Case 4: X is a hyperelliptic surface.

In this case $c_1(X)$ is a torsion element in $H_2(X,\mathbf{Z})$ and the corresponding unramified cyclic cover $\pi: X' \to X$ is an abelian surface. Every automorphism of X leaves $c_1(X)$ invariant and , hence, can be lifted to an automorphism of X'. Let L be a very ample line bundle on X. Then $\mathrm{Ker}(\mathrm{Aut}(X) \to O(S_X))$ can be identified with a subgroup of automorphisms of X' which leave invariant the class of $\pi^*(L)$ in $S_{X'}$. The argument from the previous case shows that this subgroup is algebraic. Thus,it has finitely many connected components.

Case 5: X is a ruled surface.

Let $\pi: X \to X'$ be a birational morphism onto a minimal model of X. Then, every $g \in \mathrm{Aut}(X)$ which acts identically on S_X preserves the classes of exceptional divisors of π, and hence defines an automorphism of X'. Thus $\mathrm{Ker}(\mathrm{Aut}(X) \to O(S_X))$ is isomorphic to a subgroup of Aut(X') which is a linear algebraic group.

2.THE WEYL GROUP OF AN ALGEBRAIC SURFACE. We denote by $A(X)_*$ the image $r'(A(X))$ of A(X) in $O(S_X')$. Its elements are called effective isometries of S_X'.

Now, we exhibit some isometries which (if they are defined) are obviously not effective. Let R(X) be the set of nodal curves on X , i.e. irreducible curves E on X such that

$$E^2 = -2 , E\cdot c_1(X) = 0 .$$

It follows immediately from the adjunction formula that such E is isomorphic to $\mathbb{P}^1$. It is easy to show that a curve E is nodal if and only if there exists a bimeromorphic map $f: X \to X'$ which blows down E to an ordinary double point (a node) of X'. Every $E \in R(X)$ defines an isometry of H_X

$$r_E: x \mapsto x + (x\cdot[E])[E]$$

which obviously leaves the sublattices S_X and S_X' invariant. We call it a nodal reflection. Since $r_E([E]) = -[E]$, r_E is obviously not effective.

Let W_X denote the subgroup of $O(S_X')$(or $O(S_X)$, or $O(H_X)$) generated by all nodal reflections. We call it the Weyl group of X.

Recall [B] that a Coxeter group(or system) is a pair (W,S) , where W is a group and S is a set of its generators with the defining relations

$$s^2 = 1 , (ss')^{m(s,s')} = 1 , \text{ for any } s,s' \in S,$$

where m(s,s') = ord(ss') if it is finite or ∞ otherwise.

The numbers m(s,s') can be expressed geometrically by the Coxeter graph: its vertices correspond to elements s of S and two vertices s and s' are joined by an edge only if m(s,s') > 2, the edge is labelled by m(s,s')-2 if it is greater than 1.

PROPOSITION 2. The pair $(W_X,(s_E)_{E\in R(X)})$ is a Coxeter group. Moreover,

$$\begin{aligned} m(s_E,s_{E'}) &= 1 , \text{ if } E = E', \\ &= 2, \text{ if } E\cdot E' = 0, \\ &= 3 , \text{ if } E\cdot E' = 1 \\ &= \infty , \text{ otherwise.} \end{aligned}$$

PROOF. This follows from a general result of E.Vinberg [V1] .

PROPOSITION 3. Let G be the subgroup of $O(S_X')$ generated by the subgroups $A(X)_*$ and W_X. Then W_X is a normal subgroup of G and $W_X \cap A(X)_* = \{1\}$. In particular,

$$G = A(X)_* \ltimes W_X.$$

PROOF. For every $E \in R(X)$ and $g \in A(X)$, we have

$$g_* \cdot s_E \cdot g_*^{-1} = s_{g(E)} .$$

This shows that W_X is normal in G. To prove that $W_X \cap A(X)_* = \{1\}$, we introduce the <u>fundamental chamber</u> in the vector space $V_X = S_X \otimes \mathbb{R}$:

$$C_X = \{x \in V_X : x \cdot [E] \geq 0 , \forall E \in R(X)\}.$$

Let $l: W_X \setminus \{1\} \to \mathbb{N}$ be the length function in the Weyl group W_X ($l(w)$ = the minimal number of generators s_E whose product is equal to w). If $w = s_E \cdot w'$, $l(w') < l(w)$, then

$$[E] \cdot y \leq 0 , \ \forall y \in w(C_X).$$

This follows from general properties of geometric representations of Coxeter groups (see [B],[L]). Assume that $g_* = w \in W_X \setminus \{1\}$ for some $g \in A(X)$. Write $w = s_E \cdot w'$ as above. Take $v = [H]$, where H is an ample divisor on X. Then

$$[E] \cdot g_*(v) = E \cdot g(H) \leq 0 .$$

However, g(H) is an ample divisor, hence $E \cdot g(H) > 0$. This contradiction proves the assertion.

2. AUTOMORPHISMS OF NON-RULED SURFACES. Here we restrict our attention to surfaces of non-negative Kodaira dimension (or, non-ruled surfaces). Moreover, since the automorphism group of such a surface is isomorphic to a subgroup of the automorphism group of its minimal model, we will consider only minimal surfaces. For such surfaces

$$\mathrm{Aut}(X) \cong \mathrm{Aut}(\mathbb{C}(X)/\mathbb{C}).$$

THEOREM 1. Assume that X is a minimal non-ruled surface. Then $G = A(X)_* \ltimes W_X$ is a subgroup of finite index in $O(S_X')$.

PROOF. We consider the different cases corresponding to the Enriques classification of algebraic surfaces. Let $\varkappa(X)$ denote the Kodaira dimension of X.

Case I: $\varkappa(X) = 2$ (X is of general type).

In this case $c_1(X)^2 > 0$, hence S_X' is negative definite. Hence, $O(S_X')$ is finite and the assertion is obvious.

Case II: $\varkappa(X) = 1$.

In this case, X has a unique elliptic fibration $\pi: X \to B$ induced by the map $X \to \mathbb{P}^n$, given by some multiple of the canonical linear system $|mK_X|$. In particular, $K_X^2 = c_1(X)^2 = c_X^2 = 0$ and S_X' is negative-semidefinite with the radical $\mathbb{Z}c_X$. Let $\bar{S}_X' = S_X'/\text{radical}$ and $p: S_X' \to \bar{S}_X'$ the corresponding projection. Let $\bar{S}_X'^* = \text{Hom}_{\mathbb{Z}}(\bar{S}_X', \mathbb{Z})$. For every $\varphi \in \bar{S}_X'^*$,

$$\sigma_\varphi: x \mapsto x + \varphi(p(x))c_X$$

is an isometry of S_X'. The map $\varphi \mapsto \sigma_\varphi$ defines an isomorphism of $S_X'^*$ onto a normal subgroup of $O(S_X')$ with the factor group isomorphic to the finite group $O(\bar{S}_X')$. Thus, we have an exact sequence of groups

$$1 \to \bar{S}_X'^* \to O(S_X') \to O(\bar{S}_X') \to 1 .$$

Let $S_X'(\pi)$ be the sublattice of S_X' spanned by components of fibres of π. Then $S_X'(\pi)$ is spanned by $R(X)$ and c_X. The lattice $\bar{S}_X'(\pi) = S_X'(\pi)/\mathbb{Z}c_X$ is isomorphic to the direct sum of the root lattices of type A_n, D_n or E_n. This follows either from Kodaira's classification of degenerate fibres of elliptic fibrations or from the fact that $\bar{S}_X'(\pi)$ is negative-definite and spanned by vectors v with $v^2 = -2$. Similarly to the above, we define an exact sequence

$$1 \to \bar{S}_X'(\pi)^* \to O(S_X'(\pi)) \to O(\bar{S}_X'(\pi)) \to 1.$$

It follows from the theory of affine Weyl groups ([B]) that W_X is a subgroup of finite index in $O(S_X'(\pi))$ which contains $\bar{S}_X'(\pi)^*$. In particular, W_X contains a free abelian group of rank equal to $rk(S_X'(\pi))-1$.

Let J_η be the jacobian variety of the general fibre X_η of π, then X_η is a principal homogeneous space over J_η, hence, the group of rational points of J_η over the field $\mathbb{C}(B)$ of rational function on B acts on X_η and hence on X (since X is a minimal model). It is known that $J_\eta(\mathbb{C}(B))$ is an abelian group of finite rank equal to $rk(S_X')-rk(S_X'(\pi))$, unless $J_\eta = J \otimes_{\mathbb{C}} \mathbb{C}(B)$ for some elliptic curve J over $\mathbb{C}$. In the latter case, $rk(S_X') = 1$ and $O(S_X')$ is finite. In the former case,

$$A(X)_* \supset r'(\text{Im}(J_\eta(\mathbb{C}(B)) \to A(X)).$$

It is not hard to see that $J_\eta(\mathbb{C}(B))$ maps injectively into $A(X)_*$ (compare the proof of Proposition 1 , case 2). Thus, $G = A(X)_* \ltimes W_X$ contains an extension of a free abelian group of rank $rk(S_X')-rk(S_X'(\pi))$ with help of a free abelian group of rank $rk(S_X'(\pi))-1$. It follows from the structure of $O(S_X')$ that G is of finite index in it.

Case III: $\kappa(X) = 0$. Here X belongs to one of the following classes of surfaces: a) abelian surfaces; b) K3-surfaces; c) Enriques surfaces; d) hyperelliptic surfaces.

Case IIIa): In this case $R(X) = \emptyset$, $S_X = S_X'$. We have to show that $A(X)_*$ is of finite index in $O(S_X)$. Let A be the subgroup of $O(S_X)$ of isometries which act identically on the finite discriminant group S_X^*/S_X. Clearly, A is a subgroup of finite index in $O(S_X)$ and every $\sigma \in A$ can be extended to an isometry $\tilde{\sigma}$ of H_X by checking that $\sigma \oplus 1 \in O(S_X \oplus T_X)$, where $T_X = (S_X)^{\perp}$, extends to H_X. The isometry $\tilde{\sigma}$, extended to $H^2(X,\mathbb{C})$, preserves the Hodge decomposition. By the Global Torelli Theorem for 2-tori ([BPV], Chap.V,Thm.(3.2)) $\tilde{\sigma}$ or $-\tilde{\sigma} \in A(X)_*$. This checks the assertion in this case.

Case III b): This is the most non-trivial case. The assertion is proven in [PS] as a corollary of the Global Torelli Theorem for polarized algebraic K3-surfaces.

Case III c): It was shown by V.Nikulin (unpublished) and Y.Namikawa [Na] that case III b) implies case III c). By other method it is proven in [D3].

Case III d): This is the easiest case. In this case $S_X' = S_X = H_X$ is an indefinite lattice of rank 2 which represents zero. This implies that $O(S_X')$ is finite.

COROLLARY. Let X be a minimal non-ruled surface. Then $A(X)$ is finite if and only if W_X is of finite index in $O(S_X')$.

One can interpret geometrically the condition $[O(S_X'):W_X] < \infty$ as follows. Let $V_X = S_X' \otimes \mathbb{R}$. We associate to V_X a Riemannian (resp. Euclidean, resp. Lobachevsky) space of constant curvature M^n ($n = rk(S_X') - 1$) in the usual way depending on the signature of S_X'. Then $O(S_X')$ is a discrete group of motions of this space with a fundamental domain of finite volume. Thus, W_X is of finite index in $O(S_X')$ if and only if it is a discrete subgroup of the group of motions generated by reflections into hyperplanes. Such groups are called <u>crystallographic reflection groups</u>. A fundamental domain of such group in M^n is a crystallographic polyhedron, a connected component of the complement of the union of the reflection hyperplanes in M^n (see [D2]).

4. K3-SURFACES. Recall that a K3-surface is a surface satisfying $b_1(X) = 0$ and $c_1(X) = 0$. The homology lattice $H_X \cong H_2(X,\mathbf{Z})$ is an even unimodular lattice of signature (3,19,). As such, it is isomorphic to the lattice $L_{K3} = E_8 \oplus E_8 \oplus U \oplus U \oplus U$, where U is a hyperbolic plane ($= \mathbf{Z}e_1 + \mathbf{Z}e_2$ with $e_1^2 = e_2^2 = 0$, $e_1 \cdot e_2 = 0$) and E_8 is isomorphic to the root lattice of a simple root system of type E_8 (with the values of the quadratic form multiplied by -1).

The lattice S_X may be any hyperbolic lattice which can be primitively embedded into L_{K3}. The criterions for the existence of such an embedding were given in works of M.Kneser and V.Nikulin (see [D2]).

The Weyl group W_X is equal to the reflection group $W(S_X)$ of the lattice $S_X = S_X'$, the subgroup of $O(S_X)$ generated by the reflections

$$s_\alpha : x \mapsto x + (x \cdot \alpha)\alpha$$

for all $\alpha \in S_X$ with $\alpha^2 = -2$. This easily follows from the Riemann-Roch theorem on X (see [PS]). We obtain that $\mathrm{Aut}(X) = A(X)$ is finite if and only if $[O(S_X):W(S_X)]$ is finite.

An even hyperbolic lattice is called a Nikulin lattice if its reflection group is of finite index in its orthogonal group. Thus,we have

THEOREM 2. Let X be a K3-surface. Then Aut(X) is finite if and only if S_X is a Nikulin lattice.

All Nikulin lattices were classified by V.Nikulin (the case rank = 4 was considered by E.Vinberg) in [N1] ,[N2] . It turns out that every Nikulin lattice is isomorphic to a lattice S_X for some K3-surface X (see an account of Nikulin's work in [D2]).

Note that, conjecturally, Theorem 2 holds over fields of arbitrary characteristic.

Finally, we refer to [N1] and [V2] for some more concrete results, where the description of the automorphism groups of K3-surfaces is given in terms of some infinite Coxeter groups.

5. ENRIQUES SURFACES. Recall that an Enriques surface is characterized by the conditions $b_2(X) = 10$, $2c_1(X) = 0$. This implies, in particular, that $c_1(X) \neq 0$ and it generates the torsion subgroup of $H_2(X,\mathbf{Z})$. The universal cover of X is the double cover corresponding to $c_1(X)$ and is isomorphic to a K3-surface. Conversely, if τ is

a fixed-point-free involution of a K3-surface X', then the quotient $X'/(\tau)$ is an Enriques surface.

The lattice $H_X = S_X = S_X'$ and is a unimodular even hyperbolic lattice of rank 10. As such, it is isomorphic to the lattice $L_E = E_8 \oplus U$. The latter can be also defined by the graph $T_{2,3,7}$:

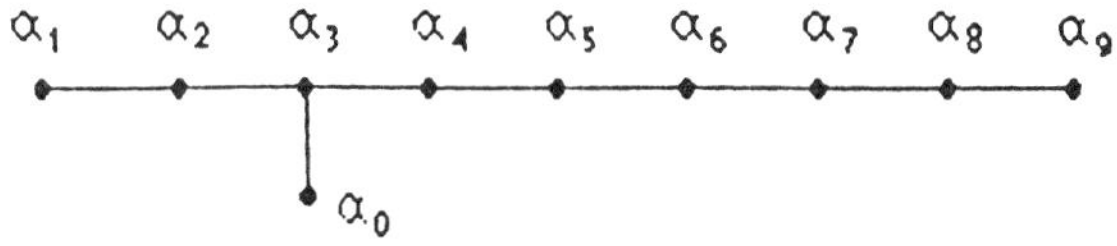

Here to each vertex of this graph is associated an element α_i (i=0,...,9) of a free base of L_E, $\alpha_i^2 = -2$ and $\alpha_i \cdot \alpha_j = 1$ or 0 dependent on whether the corresponding vertices are joined by an edge.

The orthogonal group $O(L_E) \cong W(L_E) \times \{\pm 1\}$. Also $W = W(L_E)$ is the Coxeter group with respect to the set of generators $\{s_i = s_{\alpha_i}\}_{i=0,\ldots,9}$ with the Coxeter graph equal to the above diagram.

The Weyl group W_X of an Enriques surface is "generally" trivial. The generality is understood in terms of the period space $\mathbb{P}$ for Enriques surfaces, a 10-dimensional variety parametrizing the isomorphism classes of Enriques surfaces (see [BPV]). There exists an irreducible hypersurface $\mathbb{P}^n$ in $\mathbb{P}$ such that $p \notin \mathbb{P}^n$ if and only if $R(X_p) = \emptyset$ (see [Na]).

We will call an Enriques surface <u>unnodal</u> (resp. <u>nodal</u>) if $R(X) = \emptyset$ (resp. $\neq \emptyset$).

As follows from Theorem 2, for an unnodal surface $\mathrm{Aut}(X) = A(X)$ is a subgroup of finite index in $O(S_X) \cong O(L_E)$. In particular, $A(X)$ is infinite. More precisely, the following result holds([BP],[N]):

THEOREM 3. Let X be an unnodal Enriques surface. Then $A(X)_* \subset W(H_X)$ and contains the 2-level congruence subgroup $W(H_X)(2)$ ($= \mathrm{Ker}\,(W(H_X) \to GL(H_X/2H_X))$. Moreover, there exists a countable number of hypersurfaces in $\mathbb{P}$ such that

$$\mathrm{Aut}(X) = A(X) \cong A(X)_* = W(H_X)(2) \cong W(2),$$

if X defines a point in $\mathbb{P}$ not lying on any of these hypersurfaces.

The proof relies heavily on the Global Torelli Theorem for K3-surfaces. However, one can give a proof of the first statement which works over any ground field of characteristic $\neq 2$. It uses the following purely algebraic result:

LEMMA 1. Let σ_0 be the involution of L_E defined by

$$\sigma_0 = -1_{E_8} \oplus 1_U.$$

Then $W(2)$ is the minimal normal subgroup of W containing σ_0.

This result is attributed to A.Coble [Co] , though his proof is false. A correct proof was given by E.Looijenga (unpublished).

For every embedding j of the lattice U into H_X one constructs a representation of X as a double cover of the projective plane. If g is the corresponding involution of X , then $g_* = 1_{j(U)} \oplus -1_{j(U)^\perp}$ (see [Do3]). Applying the previous lemma one sees that $A(X)_*$ contains $W(H_X)(2)$. The fact that $A(X)_* \subset W(H_X)$ is obvious ($-1 \notin A(X)_*$).

An explicit description of the automorphism group of a nodal Enriques surface is known only in a few cases. One of them is the case where X is a general nodal surface. This means that $X = X_p$, where p belongs to a Zariski open subset of $\mathbf{P}^n$. Let L_R be the sublattice of L_E given by

$$L_R = \{x \in L_E : x \cdot \alpha_9 \equiv 0 \bmod 2\}.$$

One can easily see that L_R is isomorphic to the lattice given by the graph $T_{2,4,6}$:

Let $\{\beta_0,\dots,\beta_9\}$ be a free base of L_R corresponding to the vertices of this diagram . We assume that it is ordered similarly to the case of $T_{2,3,7}$. Let W be the subgroup of $O(L_R)$ generated by the reflections $s_i = s_{\beta_i}$. The pair $(W,\{s_i\})$ is a Coxeter group. Let $W(2) = \mathrm{Ker}(W \to GL(L_R/2L_R))$ be the corresponding 2-level congruence subgroup. The quotient group $W/W(2) \cong Sp(8,\mathbf{F}_2) \ltimes (\mathbf{Z}/2)^{10}$ [Gr]. Let $\overline{W}(2)$ be the unique subgroup of W which contains $W(2)$ and $W/\overline{W}(2) \cong Sp(8,\mathbf{F}_2)$.

THEOREM 4 ([CD1]). The automorphism group of a general unnodal Enriques surface X is isomorphic to the group $\overline{W}(2)$.

The proof relies on a representation of general nodal Enriques surfaces(resp. its unversal K3-cover) as a Reye congruence in $\mathbf{P}^5$ (resp.as a quartic symmetroid surface in $\mathbf{P}^3$)(see [C]) and the following analog of Lemma 1:

LEMMA 2. Let L_i ($i = 1,2,3$) be the sublattices of L_R spanned by the vectors $\beta_0,\ldots,\beta_6$ ($i=1$), $\beta_0,\beta_2,\ldots,\beta_8$ ($i = 2$), $\beta_0,\beta_2,\ldots,\beta_7$ ($i = 3$). Then $\omega_i = -1_{L_i} \in O(L_i) = W(L_i) \subset W$ and $\overline{W}(2)$ is the minimal normal subgroup containing the elements ω_1,ω_2 and $\omega_3 \cdot s_9$.

We refer to [CD1] for the sketch of the proof and to [CD3] for complete details.

As follows from Theorem 2, Aut(X) is finite if and only if W_X is of finite index in $O(S_X)$. The first example of such a surface was suggested by G.Fano [F]. However, I still have difficulty in computing W_X in his example (his argument for the proof of the finiteness of Aut(X) is incomplete). Another example was given by the author (who was not aware of Fano's example at that time) [D3]. In this example Aut(X) $\cong D_4$, the dihedral group of order 8. The group W_X is a Coxeter group with the following Coxeter diagram

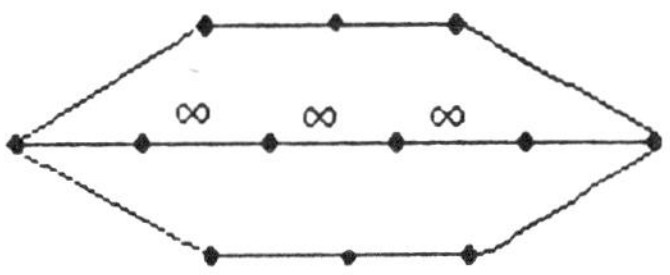

As was recently shown by V.Nikulin [N3], there are 6 classes of Enriques surfaces with finite number of automorphisms. No explicit examples,except the cited above are known.[1]

1 All Enriques surfaces with finite number of automorphisms were recently constructed by S.Kondō. Contrary to the statement of Nikulin, there are 7 classes of such surfaces.

6. RATIONAL SURFACES. In this case the assertion of Theorem 2 does not hold in general. Namely, let $X = X(\Sigma)$ be a surface obtained by blowing up a finite set $\Sigma = \{p_1,\dots,p_r\}$ $(r\geq 5)$ of points in $\mathbb{P}^2$. We have

$$H_X = S_X = \mathbb{Z}e_0+\mathbb{Z}e_1+\dots\mathbb{Z}e_r,$$

where e_0 is the class of the proper transform of a line in $\mathbb{P}^2$ and e_i is the class of the exceptional curve blown up from the point p_i $(i=1,\dots,r)$. As is well-known

$$c_X = 3e_0-e_1-\dots-e_r.$$

It is verified directly that

$$S_X' = \mathbb{Z}\alpha_0+\mathbb{Z}\alpha_1+\dots\mathbb{Z}\alpha_{r-1},$$

where $\alpha_0 = e_0-e_1-e_2-e_3, \alpha_i=e_i-e_{i+1}$, $i=1,\dots,r-1$, $\alpha_i^2 = -2$, $\alpha_i\cdot\alpha_j$ is given by the graph $T_{2,3,r-5}$ similarly to the above . In particular, we easily check that S_X' is negative definite (resp.semi-definite, resp. hyperbolic) if and only if $r \leq 8$ (resp. $r = 9$, resp. $r > 9$). If Σ is general enough (see [D1]), then $W_X = \{1\}$. If $r \geq 9$, then $O(S_X')$ is infinite, however Aut(X) is trivial (see [G],[Co1]).

In spite of the negative result above, there exist certain classes of rational surfaces $X(\Sigma)$ with $r = 9$ and 10 for which Theorem 2 holds. They correspond to special sets Σ.

$r = 9$: We let Σ be the set of base points of an Halphen pencil of plane elliptic curves of degree 3m with m-multiple points at Σ. Then, generically, $W_X = \{1\}$ (nodal curves on X correspond to irreducible components of reducible degenerate fibres of the elliptic pencil on X defined by the Halphen pencil). We have

$$O(S_X') \cong O(E_8)\ltimes\mathbb{Z}^8$$

and [G]

$$\begin{aligned} \mathrm{Aut}(X) &\cong \{1\}\ltimes(m\mathbb{Z})^8 \subset O(S_X')\,,\ m > 2 \\ &\cong \{\pm 1\}\ltimes(m\mathbb{Z})^8 \subset O(S_X')\,,\ m\leq 2. \end{aligned}$$

Note that, $O(S_X') = W$, the Coxeter group with respect to the set of the reflections s_{α_i}. If $W(m)$ is the corresponding m-level congruence subgroup, then

$$W/W(2) \cong O(E_8)/\{\pm 1\}\ltimes(\mathbb{Z}/2)^8,\quad W/W(m) \cong O(E_8)\ltimes(\mathbb{Z}/m)^8 (m \geq 2)$$

and we have

$$\mathrm{Aut}(X) \cong W(m)\,,\ m \geq 2,$$
$$\cong \overline{W}(2)\,,\ m = 1,$$

where $\overline{W}(2)$ is the unique subgroup of W which contains $W(2)$ and $W/\overline{W}(2) \cong (\mathbf{Z}/2)^8$.

$r = 10$: We let Σ to be a set of 10 points which can be realized as the nodes of an irreducible rational sextic curve . The projective isomorphism classes of such sets Σ depend on 9 parameters (instead of 12 for all sets Σ with $\#\Sigma = 10$). The corresponding surface $X(\Sigma)$ is called a <u>Coble surface</u>. It follows from above that

$$S_X' \cong L_E\,.$$

We say that a Coble surface $X = X(\Sigma)$ is <u>general</u>, if all the following properties are satisfied:

a)none of the three points of Σ are colinear;

b)none of 6 points in Σ lie on a conic;

c)none of 8 points of Σ lie on a cubic with a double point at one of them;

d) Σ does not lie on a quartic with a triple point at one of the points of Σ.

The projective isomorphism classes of such sets Σ form a quasi-projective variety $\mathfrak{C}$ of dimension 9.

PROPOSITION 4. If X is a general Coble surface, then $R(X) = \emptyset$.

The proof of this result can be given by the methods of [CD 2].

THEOREM 5. Let X be a general Coble surface. Then

$$W(L_E) \cong W(S_X')(2) \subset A(X)_* \subset W(S_X') \cong W(L_E).$$

Moreover, there exists an open Zariski subset U of $\mathfrak{C}$ such that for $\Sigma \in U$

$$\mathrm{Aut}(X(\Sigma)) \cong W(L_E)(2).$$

This result can be proven by transcendental methods (with a weaker notion of a general surface) by methods of [N1]. One can also give a purely algebraic proof valid for any sufficiently large characteristic p by the methods of [CD1]. In an equivalent form, this result is already contained in [Co2].

Notice the similarity between Theorems 3 and 5. A partial explanation of this is the fact that Coble surfaces are degenerations of Enriques surfaces.

BIBLIOGRAPHY

[AS] Algebraic surfaces (ed. by I.Shafarevich), Proc. of Steklov Math. Inst., t.75, Moscow, 1965.

[B] Bourbaki N., Groupes et algebres de Lie, Chap.IV-VI, Hermann. Paris. 1968

[BP] Barth W., Peters C., "Automorphisms of Enriques surfaces", Invent. Math.,73 (1983),383-411.

[BPV] Barth W., Peters C., Van de Ven A., Compact complex surfaces, Springer Verlag, Berlin-Heidelberg-NewYork-Tokyo, 1984.

[C] Cossec F., "Reye congruences", Trans. AMS, 280 (1983),737-751.

[CD1] Cossec F., Dolgachev I., "On automorphisms of nodal Enriques surfaces", Bul.AMS, 12, No 2 (1985), 247-249.

[CD 2] Cossec F., Dolgachev I., "Smooth rational curves on Enriques surfaces", Math.Ann. 1985(to appear).

[CD 3] Cossec F., Dolgachev I., Enriques surfaces (a book in preparation).

[Co1] Coble A., " Point sets and allied Cremona groups,II", Trans. AMS ,17 (1916), 345-385.

[Co2] Coble A., "The ten nodes of the rational sextic and of the Cayley symmetroid", Amer.J.Math., 41 (1919),243-265.

[D1] Dolgachev I., "Weyl groups and Cremona transformations", Proc. Symp. in Pure Math.,40 (1983), Part 1,283-294.

[D2] Dolgachev I., "Integral quadratic forms: applications to algebraic geometry", Sem.Bourbaki 1982/83, Asterisque, 105/106 (1984), 251-278.

[D3] Dolgachev I., "On automorphisms of Enriques surfaces", Invent. Math., 76 (1984), 163-177.

[G] Gizatullin M., "Rational G-surfaces",Izv. Akad. Nauk SSSR, 44 (1980), 110-144.

[Gr] Griess R., "Quotients of infinite reflection groups", Math.Ann., 263

(1983), 267-268.

[F] Fano G., "Superficie algebriche di genere zero e bigenere uno e loro case particulari", Rend. Circ. Matem. di Palermo, 29 (1910), 98-118.

[L] Looijenga E., "Invariant theory for generalized root systems", Invent. Math., 61 (1980), 1-32.

[M] Mumford D., Abelian varieties, Oxford Univ. Press, Oxford, 1970.

[N1] Nikulin V., "On the quotient groups of the automorphism groups of hyperbolic forms by the subgroups generated by 2-reflections", J.Soviet Math., 22 (1983), 1401-1476.

[N2] Nikulin V., "Surfaces of type K3 with finite automorphism group and the Picard group of rank 3", Proc. Steklov Math. Inst., 165 (1984), 119-230.

[N3] Nikulin V., " On the description of automorphism groups of Enriques surfaces", Dokl. Akad. Nauk SSSR, 277 (1984), 1324-1327.

[PS] Piatetsky-Shapiro I., Shafarevich I., "A Torelli theorem for algebraic surfaces of type K3", Math.USSR-Izvestia, 5 (1971), 547-588.

[V1] Vinberg E., "Discrete groups generated by reflections in Lobachevski spaces", Mat.Sbornik, 78 (1969), 471-488.

[V2] Vinberg E., "The two most algebraic K3 surfaces", Math.Ann., 265 (1983), 1-21.

DEPARTMENT OF MATHEMATICS
UNIVERSITY OF MICHIGAN
ANN ARBOR, MICHIGAN 48109
USA

Contemporary Mathematics
Volume 58, Part I, 1986

A PROBLEM ABOUT POLYNOMIAL IDEALS

Shalom Eliahou

In this paper we propose an elementary problem about polynomial ideals. It is connected to set-theoretic complete intersection problems via a lemma of Cowsik, which we recall, and which we slightly improve in the case of monomial curves.

1. The problem

Consider the polynomial algebra $A = k[x_1,\dots,x_n]$ over a field k.

Let us fix a total order $<$ on the set of monomials $x_1^{a} \dots x_n^{a_n}$ ($a_i \in \mathbb{N}$, all i) of A, compatible with the product, i.e. satisfying

$$m < n \Rightarrow mp < np$$

for all monomials m, n, p.

Given a polynomial f in A, we denote by $\max f$ its maximal monomial (for the given order $<$).

Given an ideal I in A, we denote by $\max(I)$ the ideal generated by the monomials $\max f$, for all f in I.

If I,J are two ideals, it is easily verified that

$$\max(I)\max(J) \subset \max(IJ)$$

(while the reverse inclusion may not be true). In particular,

$$\max(I^r)\max(I^s) \subset \max(I^{r+s})$$

for all r,s in $\mathbb{N}$.

Therefore the direct sum $\bigoplus_r \max(I^r)$ has a natural structure of graded algebra over A. We denote this algebra by $MAX(I)$.

We may now formulate the problem:

Problem 1.1. When is the algebra $MAX(I)$ finitely generated over A?

In other words, if r is large enough, can we always decompose $\max(I^r)$ as $\max(I^s)\max(I^t)$ for a suitable non-trivial decomposition $r = s+t$?

In section 3 we give a condition on $\max(I)$ to ensure that $MAX(I)$ be finitely generated, and give also an example where $MAX(I)$ is not finitely generated. We show that the answer depends on the order chosen.

In section 6 we prove the following result (Corollary 6.8) which we hope motivates problem 1.1 above.

Consider positive integers $a_1,\ldots,a_n$. Let C be the affine monomial curve with parametric equations $x_i = t^{a_i}$ (all i) over k. Denote by J the ideal $(t^{a_1}-x_1,\ldots,t^{a_n}-x_n)$ of $A[t]$. Order the monomials of $A[t]$ lexicographically, with $t > x_1 > \ldots > x_n$. Then we have the lemma:

Lemma. If $MAX(J)$ is finitely generated, then the curve C is a set-theoretic complete intersection in A^n.

This result is similar to and uses a beautiful lemma of R.C. Cowsik, which we recall in section 4. The author first learnt of the unpublished lemma of Cowsik from T. Vust who in turn came to know of it from M. Hochster at a meeting in Oberwolfach in June 1979.

2. Gröbner bases

Most things in this section are well-known, so almost no proofs are given. For further details, the reader may consult [1], [2], [7]. I am indebted to Gail Zacharias for the statement of proposition 2.7, and for showing me [5].

Consider the polynomial algebra $A = k[x_1,\ldots,x_n]$ over a field k.

Fix a total order on the set of monomials $\{0\} \cup \{x_1^{a} \ldots x_n^{a_n} \mid a_i \; \mathbb{N}, \text{ all } i\}$ of A, satisfying the following properties:

i) $m < n \Rightarrow mp < np$

ii) $1 \leq m$

iii) $0 < m$

for all monomials m, n, p.

Then we have the easy but crucial remark:

Remark 2.1. Any decreasing sequence of monomials eventually stops.

Given an f in A, we denote by $\max f$ its maximal monomial. Given an ideal I in A, we denote by $\max(I)$ the ideal generated by the monomials $\max f$ for all f in I.

Definition 2.2. A Gröbner basis for I is a set of elements $\{f_1,\ldots,f_r\}$ in I such that $\max f_1,\ldots,\max f_r$ generate $\max(I)$.

Using remark 2.1 above, one proves easily the following lemma:

Lemma 2.3. A Gröbner basis for I actually generates I.

The notion of Gröbner bases is very useful for various constructions on polynomial ideals. The reader may consult the references given at the beginning of this section.

Here, for convenience, we briefly recall how to detect or construct Gröbner bases for an ideal I. First some terminology.

Let f,g and $f_1,\ldots,f_r$ be in A. We say that f reduces properly to g mod$\{f_1,\ldots,f_r\}$ if we can write

$$f - a_{i1}f_{i1} - \ldots - a_{i\ell}f_{i\ell} = g$$

(where $f_{ij} \in \{f_1,\ldots,f_r\}$ and a_{ij} is a monomial with some k-coefficient, all j) with the property that the maximal monomial drops at each new term subtracted, i.e.

$$\max f > \max(f-a_{i1}f_{i1}) > \ldots > \max g.$$

Again we have an easy remark:

Remark 2.4. A set $\{f_1,\ldots,f_r\}$ is a Gröbner basis for the ideal $(f_1,\ldots,f_r)$ if and only if every f in $(f_1,\ldots,f_r)$ may be reduced properly to 0 mod $\{f_1,\ldots,f_r\}$.

Now, given two polynomials f,g with maximal monomials m,n respectively (and maximal coefficient 1 to simplify), we denote by $s(f,g)$ the polynomial

$$s(f,g) := \frac{n}{\gcd(m,n)}\, f - \frac{m}{\gcd(m,n)}\, g$$

where gcd means greatest common divisor.

The effect of this combination is to kill "minimally" the maximal monomials of f and g. Indeed $\max s(f,g) < \max f, \max g$.

The next proposition allows to decide whether a given set of polynomials is a Gröbner basis.

Proposition 2.5. A set $\{f_1,\ldots,f_r\}$ is a Gröbner basis for $(f_1,\ldots,f_r)$ if and only if for all $i \neq j$, $s(f_i,f_j)$ reduces properly to 0 mod$\{f_1,\ldots,f_r\}$.

Note: In fact it is not necessary to check all $s(f_i,f_j)$. See [2] for details.

Let us see now how to transform a given set $\{f_1,\ldots,f_r\}$ into a Gröbner basis for $(f_1,\ldots,f_r)$. Well, if $\{f_1,\ldots,f_r\}$ is not a Gröbner basis, then some $s(f_i,f_j)$ cannot be reduced properly to zero $\mathrm{mod}\{f_1,\ldots,f_r\}$. But then, reducing properly as far as possible, there will be a g in I such that

$$\left|\begin{array}{l} s(f_i,f_j) \text{ reduces properly to } g \ \mathrm{mod}\{f_1,\ldots,f_r\}; \\ \max g \text{ does not belong to } (\max f_1,\ldots,\max f_r). \end{array}\right.$$

In that case one should replace $\{f_1,\ldots,f_r\}$ by $\{f_1,\ldots,f_r,g\}$.

After a finite number of such transformations, the process ends up with a Gröbner basis $\{f_1,\ldots,f_r,g_1,\ldots,g_s\}$ for $(f_1,\ldots,f_r)$.

We give now two examples of application of Gröbner bases, which will be used in section 6. Namely, computing the homogenization of a given ideal and, respectively, its intersection with the polynomial algebra in one less variable.

Let t be a new variable. Given an f in A, we define its homogenization f_* as usual, as

$$f_* := t^{\deg f} f\left(\frac{x_1}{t},\ldots,\frac{x_n}{t}\right) \in A[t].$$

Given an ideal I in A, we define its homogenization I_* as $I_* :=$ ideal in $A[t]$ generated by all f_*, with f in I.

Proposition 2.6. Let $\{f_1,\ldots,f_r\}$ be a Gröbner basis for I with respect to any ordering of monomials satisfying i), ii), iii) and

iv) $\deg m < \deg n \Rightarrow m < n$ (all monomials m,n).

Then I_* is generated by $f_{1*},\ldots,f_{r*}$.

The proof consists in considering $\max(I_*)$ and using lemma 2.3. See e.g. [7].

Proposition 2.7. Let J be an ideal in $A[t]$, and let $\{g_1,\ldots,g_r\}$ be a Gröbner basis for J with respect to the lexicographic order induced by $t > x_1 > \ldots > x_n$. Then $J \cap A$ is generated by $\{g_1,\ldots,g_r\} \cap A$.

Again, the proof consists in considering $\max(J \cap A)$ and using lemma 2.3.

We close this section with an observation that will be used in section 3.

Lemma 2.8. Let f,g in A be such that $\max f$, $\max g$ have no common factor (equivalently form a regular sequence). Then $s(f,g)$ reduces properly to $0 \bmod\{f,g\}$.

Proof. Write $f = \sum_{i=1}^{r} m_i$, $g = \sum_{j=1}^{s} n_j$, where m_i, n_j are monomials with some k-coefficient, ordered decreasingly. By hypothesis,

$$s(f,g) = n_1 f - m_1 g.$$

From $gf - fg = 0$ we obtain

$$s(f,g) + \left[\sum_{2}^{s} n_i f - \sum_{2}^{r} m_i g \right] = 0.$$

But now it is easy to see that the terms in the square bracket may be permuted so as to produce a proper reduction of $s(f,g)$ to $0 \bmod\{f,g\}$. ∎

Corollary 2.9. Let $f_1,\dots,f_r$ in A be such that $\max f_1,\dots,\max f_r$ form a regular sequence. Then $\{f_1,\dots,f_r\}$ is a Gröbner basis for $(f_1,\dots,f_r)$.

Proof. The hypothesis amounts to say that for all $i \neq j$, $\max f_i$ and $\max f_j$ have no common factor. Now apply lemma 2.8 and proposition 2.5. ∎

3. Examples

Consider the polynomial algebra $A = k[x_1,\dots,x_n]$ over a field k. Fix a total order on the set of monomials of A satisfying the same properties as in section 2.

Let I be an ideal in A. Define $MAX(I) = \bigoplus_r \max(I^r)$. Our problem is: when is $MAX(I)$ finitely generated as an A-algebra?

We first give a condition on $\max(I)$ implying finite generation of $MAX(I)$. Then we give an example where $MAX(I)$ is not finitely generated. Finally we show that the answer depends on the order chosen.

Proposition 3.1. If $\max(I)$ is generated by a regular sequence then $\max(I^r) = \max(I)^r$, all r. In other words $MAX(I)$ is finitely generated over A, by $\max(I)$.

Proof. Let $f_1,\dots,f_s$ in I be such that their maximal monomials form a regular sequence and generate $\max(I)$.

We will show by induction on r that the set $\{f_1,\dots,f_s\}^r$ (notation self-explanatory) is a Gröbner basis for I^r, using proposition 2.5.

By construction it is true for $r = 1$. Assume now it is true for $r-1$. Let $f = \prod_i f_i^{a_i}$, $g = \prod_i f_i^{b_i}$ $(\Sigma a_i = \Sigma b_i = r)$ be any pair of elements in $\{f_1,\dots,f_s\}^r$.

i) If f and g have some factor f_i in common, write $f' = ff_i^{-1}$ and $g' = gf_i^{-1}$. Then f',g' belong to I^{r-1}, and since $\{f_1,\dots,f_s\}^{r-1}$ is a Gröbner basis for I^{r-1}, $s(f',g')$ reduces properly to $0 \bmod \{f_1,\dots,f_s\}^{r-1}$. Therefore $s(f,g) = f_i s(f',g')$ reduces properly to $0 \bmod \{f_1,\dots,f_s\}^r$.

ii) If f and g have no factor f_i in common, then $\max f$ and $\max g$ have no common factor either, since the $\max f_i$'s have no common factor pairwise. By lemma 2.8, $s(f,g)$ reduces properly to $0 \bmod \{f_1,\dots,f_s\}^r$.

The desired result follows from i), ii) and proposition 2.5.∎

Now we give an example of an ideal I in $k[x,y]$ such that $MAX(I)$ is not finitely generated.

Example 3.2. Let $I = (x^2+y^2, xy)$. We consider the lexicographic order with $x > y$.

For convenience, let us introduce a new variable t and write $MAX(I)$ as $\bigoplus_r \max(I^r)t^r$. We will prove that $MAX(I)$ is not finitely generated.

Observation. A Gröbner basis for I is $\{x^2+y^2, xy, y^3\}$ (notice that $y^3 = s(x^2+y^2, xy)$). This is immediate using proposition 2.5.

Claim 1. $y^{2r+1}t^r$ belongs to $MAX(I)$, all $r \geq 1$. Actually y^{2r+1} belongs to I^r itself (hence also to $\max(I^r)$) as we see by induction on r:

For $r = 0$, $y \in I^0 = A$. For $r = 1$, $y^3 \in I$.

Now suppose $r \geq 2$, $y^{2r-3} \in I^{r-2}$ and $y^{2r-1} \in I^{r-1}$.

Then $y^{2r+1} = (x^2+y^2)y^{2r-1} - (xy)^2y^{2r-3} \in I^r$.

Claim 2. $y^{2r}t^r$ does not belong to MAX(I), all $r \geq 1$.

We must show that no f in I^r has y^{2r} as maximal monomial. Among the canonical generators of I^r, namely the elements of $\{x^2+y^2,xy\}^r$, the only one having y^{2r} as monomial (not necessarily maximal) is $(x^2+y^2)^r$, which turns out to be also the only one having x^{2r} as monomial; and x^{2r} is bigger than y^{2r}. Therefore if y^{2r} arises as monomial of a given f in I^r, then x^{2r} does too. Hence y^{2r} cannot be max f.

As a consequence of claim 2, $y^{2r+1}t^r$ cannot be a polynomial expression in elements of lower degree of MAX(I). Therefore MAX(I) is not finitely generated. ∎

Now it is easy to derive an example of an ideal I for which MAX(I) is or is not finitely generated according to the order chosen.

Example 3.3. Let $I = (x^2+y^2,xy+z)$ in $k[x,y,z]$.

Then for the lexicographic order induced by $x > y > z$, MAX(I) is not finitely generated. This follows easily from example 3.2.

Now take the lexicographic order induced by $z > x > y$. Then a Gröbner basis for I is $\{x^2+y^2,z+xy\}$ since x^2,z is a regular sequence (corollary 2.9), and hence MAX(I) is finitely generated by proposition 3.1. ∎

Slightly modifying this example one gets a more "natural" example.

Example 3.4. Let $P \subset k[x,y,z]$ be the ideal of the monomial curve (t^2,t^3,t^5) in A^3. Then $P = (x^3-y^2,xy-z)$, and the assertions are the same as before:

MAX(P) is not finitely generated for the lexicographic order induced by $x > y > z$, while MAX(P) is finitely generated for the one induced by $z > x > y$. ∎

4. Cowsik's lemma

Consider a commutative noetherian ring A, and P a prime ideal in A. Let us denote by $P^{(r)}$ the r-th symbolic power of P, i.e. the P-primary component of P^r, or more concretely the ideal $\{x \in A | sx \in P^r$, some s in $A \setminus P\}$.

The formula $P^{(r)}P^{(s)} \subset P^{(r+s)}$ is easily verified, so the direct sum $\bigoplus_r P^{(r)}$ has a natural structure of graded A-algebra. We denote this algebra by S(P).

Now suppose furthermore that $A = (A,M)$ is local, of dimension n, with residue field infinite, and that P is distinct from M.

<u>Lemma</u> 4.1 (Cowsik). If $S(P)$ is finitely generated as an A-algebra, then there exist $n-1$ elements $b_1,\ldots,b_{n-1}$ in P such that

$$P = \mathrm{rad}(b_1,\ldots,b_{n-1}).$$

<u>Note</u>. The lemma is still valid in case P is not prime and does not have M as minimal prime. In the non-prime case $P^{(r)}$ is defined as $(P^r \cdot S^{-1}A) \cap A$, where S is the intersection of the complements of the minimal primes of P.

Let us quote first a standard lemma.

<u>Lemma</u> 4.2. Let $B = \bigoplus_r B_r$ be a finitely generated graded algebra over a noetherian ring B_0. Then there exists an integer N such that the subalgebra $B^{(N)} := \bigoplus_r B_{rN}$ is generated over B_0 by B_N, i.e. $B_{rN} = (B_N)^r$, all r.

<u>Proof of</u> 4.1. Let N be such that $P^{(rN)} = P^{(N)^r}$, all r. (lemma 4.2)

Call $Q := P^{(N)}$.

The Krull dimension of $\bigoplus_r Q^r/Q^{r+1}$ is n. Since all powers of Q are P-primary, and $P \neq M$, any element of $M/Q - P/Q$ is a non zero-divisor. Therefore the Krull dimension of $\bigoplus_r Q^r/MQ^r$ is at most $n-1$. But then one can apply Noether's normalization lemma: there exist $n-1$ elements $\bar{b}_1,\ldots,\bar{b}_{n-1}$ in Q/MQ such that the ring extension

$$A/M[\bar{b}_1,\ldots,\bar{b}_{n-1}] \subset \bigoplus_r Q^r/MQ^r$$

is integral.

Now let $b_1,\dots,b_{n-1} \in Q$ be lifts of $\bar{b}_1,\dots,\bar{b}_{n-1}$ respectively. The fact that the ring extension above is integral implies that

$$Q^s \subset (b_1,\dots,b_{n-1}) + MQ^s$$

for some large number s.

By Nakayama's lemma,

$$Q^s \subset (b_1,\dots,b_{n-1})$$

hence $P = \text{rad } Q = \text{rad}(b_1,\dots,b_{n-1})$. ∎

Now in case A is the polynomial algebra $k[x_1,\dots,x_n]$ over an algebraically closed field k, and P is a height $n-1$ prime ideal, i.e. P defines a curve C in A^n, Cowsik's lemma translates as follows:

Lemma 4.3 (Cowsik). If $S(P)$ is finitely generated, then C is locally a set-theoretic complete intersection in A^n, i.e. is locally the intersection of $n-1$ hypersurfaces.

To determine when $S(P)$ is finitely generated seems to be a rather difficult problem. Very recently, Cowsik constructed an example of a polynomial prime ideal P in 3 variables where $S(P)$ is not finitely generated (private communication).

On the other hand, $S(P)$ was proved to be finitely generated in certain cases (see [3], [4], [6]).

5. The case of monomial curves

Let $a_1,\dots,a_n$ be positive integers, and C the affine monomial curve of parametric equations $x_i = t^{a_i}$ (all i) over an algebraically closed field k.

Let $P \subset A = k[x_1,\dots,x_n]$ be the ideal of C. Then we have the following refinement of Cowsik's lemma 4.3 in this case.

Lemma 5.1. If $S(P)$ is finitely generated, then C is globally a set-theoretic complete intersection in A^n.

The proof reproduces the one in the general case, except that here we take into account the homogeneity of P and its symbolic powers for a suitable graduation on A.

Indeed, let us give to A the graduation induced by $\deg x_i := a_i$, all i. Then P clearly becomes homogeneous. Also all $P^{(r)}$ are homogeneous; this follows from primary decomposition in the homogeneous case.

Proof of 5.1. As before let N be such that $P^{(rN)} = P^{(N)r}$, all r (lemma 4.2).

Call $Q := P^{(N)}$.

Let M be the maximal ideal $(x_1,\ldots,x_n)$, which of course contains P. Again, since all powers of Q are P-primary, the Krull dimension of the algebra $\bigoplus_r Q^r/MQ^r$ is at most $n-1$.

Let $q_1,\ldots,q_s$ be homogeneous generators of Q, of degree $d_1,\ldots,d_s$ respectively, and let $\bar{q}_i$ denote the image of q_i in Q/MQ. Then

$$\bigoplus_r Q^r/MQ^r = k[\bar{q}_1,\ldots,\bar{q}_s].$$

Now let $d = \Sigma d_i$, and define $\bar{p}_i$ as $\bar{p}_i = \bar{q}_i^{\,d-d_i}$, all i. Then $\bar{p}_1,\ldots,\bar{p}_s$ are homogeneous of the same degree d, and the ring extension

$$k[\bar{p}_1,\ldots,\bar{p}_s] \subset k[\bar{q}_1,\ldots,\bar{q}_s]$$

is integral. Therefore both rings have the same dimension, which is at most $n-1$.

By Noether's normalization lemma, there exist $n-1$ elements $\bar{b}_1,\ldots,\bar{b}_{n-1}$, linear in the $\bar{p}_i$'s, such that the ring extension

$$k[\bar{b}_1,\ldots,\bar{b}_{n-1}] \subset k[\bar{p}_1,\ldots,\bar{p}_s]$$

is integral, hence also the ring extension

$$k[\bar{b}_1,\ldots,\bar{b}_{n-1}] \subset k[\bar{q}_1,\ldots,\bar{q}_s]. \qquad (*)$$

Now, by construction of the $\bar{b}_i$'s, we can find $b_1,\ldots,b_{n-1}$ in Q, homogeneous of degree d, which lift $\bar{b}_1,\ldots,\bar{b}_{n-1}$ respectively.

The fact that the extension (*) is integral implies that

$$Q^t \subset (b_1,\ldots,b_{n-1}) + MQ^t$$

for some large number t. Iterating, we get

$$Q^t \subset (b_1,\ldots,b_{n-1}) + M^m Q^t$$

for m arbitrarily large. Since Q^t, M^m and, most importantly, $(b_1,\ldots,b_{n-1})$ are homogeneous ideals, we deduce that

$$Q^t \subset (b_1,\ldots,b_{n-1}).$$

Hence $P = \mathrm{rad}\ Q = \mathrm{rad}(b_1,\ldots,b_{n-1})$.∎

Remark 5.2. In [4], the monomial curve (t^4,t^6,t^7,t^9) is proved to be a set-theoretic complete intersection in A^4, via lemma 5.1.

6. Maximal monomials and symbolic powers

Here we exhibit a connection between problem 1.1 and the problem of finite generation of $S(P)$, when P defines a monomial curve in affine n-space. Similar and more direct connections probably exist for any polynomial (prime) ideal P.

Consider the polynomial algebra $A = k[x_1,\ldots,x_n]$ over a field k. Let $a_1,\ldots,a_n$ be positive integers and $C \subset A^n$ the monomial curve of parametric equations $x_i = t^{a_i}$, all i. Let $P \subset A$ be the ideal of C. Consider the ideal

$$J = (t^{a_1}-x_1,\ldots,t^{a_n}-x_n) \quad \text{in} \quad A[t].$$

Order the monomials of $A[t]$ lexicographically, with $t > x_1 > \ldots > x_n$.

The claimed connection is given by the following lemma.

Lemma 6.1. If $MAX(J)$ is finitely generated then $S(P)$ is finitely generated.

The proof is in two steps, the second one being the equality $P^{(r)} = J^r \cap A$. The first step is the following general preposition.

Proposition 6.2. Let G be an ideal in $A[t]$. If $MAX(G)$ in finitely generated (order: lexicographic with $t > x_1 > \dots > x_n$), then the graded algebra

$$\bigoplus_r (G^r \cap A)$$

is finitely generated.

Proof. By lemma 4.2, there is a positive integer N such that

$$\max(G^{rN}) = \max(G^N)^r, \quad \text{all } r.$$

This means the following. If $\{g_1,\dots,g_s\}$ is a Gröbner basis for G^N, then $\{g_1,\dots,g_s\}^r$ is a Gröbner basis for G^{rN}.

Now by proposition 2.7, $G^{rN} \cap A$ is generated by $\{g_1,\dots,g_s\}^r \cap A$, which clearly is the same set as $(\{g_1,\dots,g_s\} \cap A)^r$. Therefore

$$G^{rN} \cap A = (G^N \cap A)^r, \quad \text{all } r.$$

Now the algebra $\bigoplus_r (G^r \cap A)$ is module-finite over the subalgebra $\bigoplus_r (G^{rN} \cap A)$ which is finitely generated by $G^N \cap A$ as we just saw. Therefore $\bigoplus_r (G^r \cap A)$ is finitely generated.∎

Let us return to P = the ideal of the monomial curve C and J = the ideal $(t^{a_1}-x_1,\dots,t^{a_n}-x_n)$. We will show that $P^{(r)} = J^r \cap A$. First some preliminaries.

Consider $A = k[x_1,\dots,x_n]$ with the graduation induced by $\deg x_i := a_i$, all i. Then P is homogeneous in A.

Now let $M \subset A$ be the ideal $(x_1-1,\dots,x_n-1)$. Of course $P \subset M$ since the point $(1,\dots,1)$ lies on C. We will relate $P^{(r)}$ with M^r.

Notation 6.3. For any ideal I in A, denote by $h(I)$ the largest homogeneous subideal of I.

Lemma 6.4. $P^{(r)} = h(M^r)$, all r.

Proof. See [4].∎

So we must understand better what is $h(I)$ for a given ideal I in A. Let us denote by I_* the homogenization of I in $A[t]$ (see section 2).

Lemma 6.5. $h(I) = I_* \cap A$.

Proof. Clearly $h(I)$ is contained in $I_* \cap A$. For the converse, first recall (or observe) that if $f(t,x_1,\ldots,x_n)$ belongs to I_*, then

$$f(t = 1) := f(1,x_1,\ldots,x_n)$$

belongs to I. Now $I_* \cap A$ is homogeneous, therefore contained in $h(I)$ if contained in I. But this is the case: if g belongs to $I_* \cap A$, then $g = g(t=1)$ and hence g belongs to I.∎

Let us return to $M = (x_1-1,\ldots,x_n-1)$.

Lemma 6.6. $M_* = (x_1-t^{a_1},\ldots,x_n-t^{a_n})$.

Proof. By proposition 2.6, one obtains generators for M_* by simply homogenizing a Gröbner basis for M with respect to a suitable ordering of the monomials. But $\{x_1-1,\ldots,x_n-1\}$ is a Gröbner basis with respect to any ordering since $x_1,\ldots,x_n$ is a regular sequence (by corollary 2.9; recall also that monomials are ≥ 1). The lemma is proved, remembering that the x_i's are of degree a_i.∎

Now recall (or observe) that the homogenization $*$ commutes with product. So let's denote by M_*^r the ideal $(M_*)^r = (M^r)_*$.

Lemma 6.7. $M_*^r \cap A = P^{(r)}$, all r.

Proof. By lemmas 6.4 and 6.5.∎

Proof of 6.1. Observe that "J" is another name for M_*. Apply proposition 6.2 and lemma 6.7.∎

Corollary 6.8. If MAX(J) is finitely generated then C is a set-theoretic complete intersection in A^n.

Proof. By lemma 6.1 and lemma 5.1.∎

Acknowledgements

I am indebted to R.C. Cowsik, S. Gitler, M. Kervaire, N. Mohan Kumar, T. Vust and G. Zacharias for many fruitful conversations concerning this work.

References

[1] Buchberger B.: A theoretical basis for the reduction of polynomials to canonical form. ACM SIGSAM Bull., Vol. 10, No. 4, (1976), 19-29.

[2] Buchberger B.: A criterion for detecting unnecessary reductions in the construction of Gröbner bases. Proc. EUROSAM 79, Lect. Notes Comp. Sc. 72, Springer, New York (1979).

[3] De Concini C., Eisenbud D. and Procesi C.: Young diagrams and determinantal varieties. Inv. Math. 56 (1980), 129-165.

[4] Eliahou S.: Symbolic powers of monomial curves. Preprint (1985).

[5] Gianni P., Trager B. and Zacharias G.: Primary decomposition of polynomial ideals via Gröbner bases. Draft of a preprint (1984).

[6] Huneke C.: On the finite generation of symbolic blow-ups. Math. Z. 179 (1982), 465-472.

[7] Robbiano L. and Valla G.: On set-theoretic complete intersections in the projective space. Preprint.

Departamento de Matemáticas
Centro de Investigación y de Estudios Avanzados
Apartado Postal 14-740
México, D. F., C.P. 07000

Contemporary Mathematics
Volume 58, Part I, 1986

LOCAL TORELLI THEOREM FOR CERTAIN EXTREMAL VARIETIES.

Carlos Gómez-Mont Avalos

In this note we announce the result that the local Torelli theorem holds for certain extremal varieties in complex projective space. Details will appear in [3]. I would like to thank my advisor Phillip Griffiths.

Let $\mathbb{P}^n$ be the complex projective space of dimension n. Let V denote a smooth nondegenerate subvariety of dimension k and degree d in $\mathbb{P}^n$. Let $p_g(V)$ denote the geometric genus of V. Then we have the following theorem of Harris (reference [2]).

Theorem (J.Harris): The greatest possible value of $p_g(V)$ is

$$\binom{M}{k+1}(n-k) + \binom{M}{k}(d-1-M(n-k))$$

where $M = \left[\frac{d-1}{n-k}\right]$ and ( ) represents binomial coefficient.

A variety V for which $p_g(V)$ has the maximum value given by the above theorem is called an *extremal variety*. The degree of V is assumed to be $\geq k(n-k)+2$

If $(n-k) \geq 2$ and $d \geq 2n-2k+3$, then it follows from theorem (3.17) of P. Griffiths and J. Harris (reference [1]) that an extremal V can be realized as a hypersurface in a variety of dimension $(k+1)$ of minimal degree $(n-k)$.

Construction and classification of extremal varieties in $\mathbb{P}^n$ is carried out in [2]. Consider the Chow scheme of cycles of dimension k and degree d in $\mathbb{P}^n$. Let E denote the open subset comprising the extremal varieties.

Let

$$M = [(d-1)/(n-k)]$$

$$A_1 = d-(M+1)(n-k)$$

$$A_2 = ((k-2)(d-(M+1)(n-k))+(n-k-2))-1$$

$$A_3 = (k-2)(M+1)-(M+1)$$

$$A_4 = (-k(d-(M+1)(n-k))$$

$$A_5 = 2 + k(M+1)-n-k$$

$$A_6 = (k-2)(d-(M+1)(n-k))-n+k$$
$$A_7 = (2-k)(M+1)+k-1$$
$$A_8 = (k-1)(d-(M+1)(n-k))-2(n-k)+4$$
$$A_9 = -(k-1)(M+1)$$
$$A_{10} = -n+k-(K+1)(d-M(n-k)-2)$$
$$A_{11} = (k+1)(M-k)-(k+1)$$
$$A_{12} = -2(n-k-2)-k(d-M(n-k)-2)-2$$
$$A_{13} = k(M-k)-2(k+1)$$

Then we have the following;

Theorem: For $k \neq 1,2,3, n-4, n-2, n-1$ and $d \geq 3.\,k.\,(n-k)+3$,

If $A_1 \geq -M; A_2 < A_3,\ A_4 < A_5,$ $A_6 > A_7,\ A_8 > A_9,\ A_{10} < A_{11}$ and $A_{12} < A_{13}$ then for the class E the local Torelli theorem holds.

Remarks: The condition on k is necessary to avoid certain exceptional extremals. The condition $A_1 \geq -M$ is necessary for considering an extremal variety as an ample divisor in the minimal variety. The conditions $A_2 < A_3,\ A_4 < A_5,\ A_6 > A_7,\ A_8 > A_9$ are necessary for annihilating some of the groups appearing in the spectral sequence for calculating the cohomology of extremal varieties. The conditions $A_{10} < A_{11},\ A_{12} < A_{13}$ are needed for certain calculations.

We have checked by computer that the various conditions mentioned in the theorem are satisfied in $\mathbb{P}^9$ for $k = 4$ and $63 \leq d \leq 77$, in $\mathbb{P}^{10}$ for $k = 4$ and $75 \leq d \leq 89$, and for $k = 5$ and $78 \leq d \leq 91$

```
N =   9

9   4   63      12  -5    3    13   -2    8     1        -2  >=?-12  YES
              0 <? 21  YES           8 <?   49    YES        -9  >?-23  YES
            -12 >?-39  YES          -10   <?   35 YES         -12 <?  22 YES

9   4   64      12  -5    3    13   -1    8     2        -1  >=?-12  YES
             -2 <? 21  YES           4 <?   49    YES        -7  >?-23  YES
             -9 >?-39  YES         -15 <?   35    YES       -16   <?  22 YES

9   4   65      12  -5    3    13    0    8     3         0  >=?-12  YES
             -4 <? 21  YES           0 <?   49    YES        -5  >?-23  YES
             -6 >?-39  YES         -20   <? 35    YES       -20   <?  22 YES

9   4   66      13  -5    3    14   -4    9    -1        -4  >=?-13  YES
              4 <? 23  YES          16    <?   53 YES        -13 >?-25  YES
            -18 >?-42  YES           0    <?   40 YES        -4 <? 26   YES

9   4   67      13  -5    3    14   -3    9     0        -3  >=?-13  YES
              2 <? 23  YES          12    <?   53 YES        -11 >?-25  YES
            -15 >?-42  YES          -5    <?   40 YES        -8 <? 26   YES

9   4   68      13  -5    3    14   -2    9     1        -2  >=?-13  YES
              0 <? 23  YES          8    <?   53 YES       -9 >?-25  YES
            -12 >?-42  YES          -10      <? 40 YES       -12 <?  26 YES

9   4   69      13  -5    3    14   -1    9     2        -1  >=?-13  YES
             -2 <? 23  YES          4    <?    53 YES       -7 >?-25  YES
             -9 >?-42  YES         -15    <?   40 YES       -16 <?  26  YES


9   4   70      13  -5    3    14    0    9     3         0  >=?-13  YES
             -4 <? 23  YES           0   <?   53 YES        -5 >?-25  YES
             -6 >?-42  YES         -20    <?   40 YES        -20 <? 26  YES

9   4   71      14  -5    3    15   -4    10   -1        -4  >=?-14  YES
              4 <? 25  YES          16    <?   57 YES        -13  >?-27  YES
            -18 >?-45  YES           0    <?   45 YES        -4  <? 30  YES

9   4   72      14  -5    3    15   -3    10    0        -3  >=?-14  YES
              2 <? 25  YES          12    <?   57 YES        -11  >?-27  YES
            -15 >?-45  YES          -5    <?   45 YES        -8  <? 30  YES

9   4   73      14  -5    3    15   -2    10    1        -2  >=?-14  YES
              0 <? 25  YES          8    <? 57    YES        -9  >?-27  YES
            -12 >?-45  YES          -10   <?   45 YES        -12  <?  30 YES

9   4   74      14  -5    3    15   -1    10    2        -1  >=?-14  YES
             -2 <? 25  YES          4    <? 57    YES        -7  >?-27  YES
             -9 >?-45  YES          -15 <? 45    YES        -16  <?  30 YES

9   4   75      14  -5    3    15    0    10    3         0  >=?-14  YES
             -4 <? 25  YES          0    <? 57    YES        -5  >?-27  YES
             -6 >?-45  YES          -20 <? 45    YES        -20  <?  30 YES
```

```
 9    4    76        15   -5    3   16    -4   11   -1        -4 >=?-15  YES
                 4  <? 27   YES             16   <?  61  YES     -13 >?-29  YES
               -18  >?-48   YES              0   <?  50  YES      -4 <? 34  YES

 9    4    77        15   -5    3   16    -3   11    0        -3 >=?-15  YES
                 2  <? 27   YES             12   <?  61  YES     -11 >?-29  YES
               -15  >?-48   YES             -5   <?  50  YES      -8 <? 34  YES


N     =    10


10    4    75        12   -5    4   13    -3    8    1        -3  >=?-12  YES
                 1  <? 21   YES             12 <?    48  YES     -12  >?-23  YES
               -17  >?-39   YES            -11  <?   35  YES      -14 <?  22  YES

10    4    76        12   -5    4   13    -2    8    2        -2  >=?-12  YES
                -1  <? 21   YES              8  <?   48  YES     -10  >?-23  YES
               -14  >?-39   YES            -16  <?   35  YES      -18  <?  22  YES

10    4    77        12   -5    4   13    -1    8    3        -1  >=?-12  YES
                -3  <? 21   YES            4  <?  48  YES         -8  >?-23  YES
               -11  >?-39   YES           -21   <? 35  YES        -22   <?  22  YES

10    4    78        12   -5    4   13     0    8    4         0  >=?-12  YES
                -5  <? 21   YES           0 <?   48  YES          -6  >?-23  YES
                -8  >?-39   YES           -26 <? 35  YES          -26 <? 22  YES


10    4    79        13   -5    4   14    -5    9   -1        -5  >=?-13  YES
                 5  <? 23   YES             20 <?  52  YES       -16 >?-25  YES
               -23  >?-42   YES             -1 <?  40  YES       -6 <?  26  YES

10    4    80        13   -5    4   14    -4    9    0        -4  >=?-13  YES
                 3  <? 23   YES           16  <?  52  YES        -14 >?-25  YES
               -20  >?-42   YES           -6  <?  40  YES        -10 <? 26  YES

10    4    81        13   -5    4   14    -3    9    1        -3  >=?-13  YES
                 1  <? 23   YES           12  <?  52  YES        -12 >?-25  YES
               -17  >?-42   YES             -11 <? 40  YES         -14 <? 26  YES

10    4    82        13   -5    4   14    -2    9    2        -2  >=?-13  YES
                -1  <? 23   YES           8  <?  52   YES        -10  >?-25  YES
               -14  >?-42   YES           -16 <?  40  YES          -18 <? 26  YES

10    4    83        13   -5    4   14    -1    9    3        -1  >=?-13  YES
                -3  <? 23   YES           4  <?  52   YES        -8   >?-25  YES
               -11  >?-42   YES           -21 <?  40  YES         -22  <? 26 YES

10    4    84        13   -5    4   14     0    9    4         0  >=?-13  YES
                -5  <? 23   YES           0  <?  52   YES        -6   >?-25  YES
                -8  >?-42   YES         -26  <? 40  YES           -26  <? 26 YES
```

10	4	85	14	-5	4	15	-5	10	-1	-5 >=?-14 YES
			5 <? 25	YES			20 <? 56		YES	-16 >?-27 YES
			-23 >?-45	YES			-1 <? 45		YES	-6 <? 30 YES
10	4	86	14	-5	4	15	-4	10	0	-4 >=?-14 YES
			3 <? 25	YES			16 <? 56		YES	-14 >?-27 YES
			-20 >?-45	YES			-6 <? 45		YES	-10 <? 30 YES
10	4	87	14	-5	4	15	-3	10	1	-3 >=?-14 YES
			1 <? 25	YES			12 <? 56		YES	-12 >?-27 YES
			-17 >?-45	YES			-11 <? 45		YES	-14 <? 30 YES
10	4	88	14	-5	4	15	-2	10	2	-2 >=?-14 YES
			-1 <? 25	YES			8 <? 56		YES	-10 >?-27 YES
			-14 >?-45	YES			-16 <? 45		YES	-18 <? 30 YES
10	4	89	14	-5	4	15	-1	10	3	-1 >=?-14 YES
			-3 <? 25	YES			4 <? 56		YES	-8 >?-27 YES
			-11 >?-45	YES			-21 <? 45		YES	-22 <? 30 YES

10	5	78	15	-6	3	16	-2	10	1	-2 >=?-15 YES
			2 <? 42	YES			10 <? 77		YES	-11 >?-44 YES
			-14 >?-64	YES			-11 <? 54		YES	-13 <? 38 YES
10	5	79	15	-6	3	16	-1	10	2	-1 >=?-15 YES
			-1 <? 42	YES			5 <? 77		YES	-8 >?-44 YES
			-10 >?-64	YES			-17 <? 54		YES	-18 <? 38 YES
10	5	80	15	-6	3	16	0	10	3	0 >=?-15 YES
			-4 <? 42	YES			0 <? 77		YES	-5 >?-44 YES
			-6 >?-64	YES			-23 <? 54		YES	-23 <? 38 YES
10	5	81	16	-6	3	17	-4	11	-1	-4 >=?-16 YES
			8 <? 45	YES			20 <? 82		YES	-17 >?-47 YES
			-22 >?-68	YES			1 <? 60		YES	-3 <? 43 YES
10	5	82	16	-6	3	17	-3	11	0	-3 >=?-16 YES
			5 <? 45	YES			15 <? 82		YES	-14 >?-47 YES
			-18 >?-68	YES			-5 <? 60		YES	-8 <? 43 YES
10	5	83	16	-6	3	17	-2	11	1	-2 >=?-16 YES
			2 <? 45	YES			10 <? 82		YES	-11 >?-47 YES
			-14 >?-68	YES			-11 <? 60		YES	-13 <? 43 YES
10	5	84	16	-6	3	17	-1	11	2	-1 >=?-16 YES
			-1 <? 45	YES			5 <? 82		YES	-8 >?-47 YES
			-10 >?-68	YES			-17 <? 60		YES	-18 <? 43 YES
10	5	85	16	-6	3	17	0	11	3	0 >=?-16 YES
			-4 <? 45	YES			0 <? 82		YES	-5 >?-47 YES
			-6 >?-68	YES			-23 <? 60		YES	-23 <? 43 YES

10	5	86	17 -6	3	18	-4 12 -1	-4 >=?-17 YES
			8 <? 48 YES			20 <? 87 YES	-17 >?-50 YES
			-22 >?-72 YES			1 <? 66 YES	-3 <? 48 YES
10	5	87	17 -6	3	18	-3 12 0	-3 >=?-17 YES
			5 <? 48 YES			15 <? 87 YES	-14>?-50 YES
			-18 >?-72 YES			-5 <? 66 YES	-8 <? 48 YES
10	5	88	17 -6	3	18	-2 12 1	-2 >=?-17 YES
			2 <? 48 YES			10 <? 87 YES	-11 >?-50 YES
			-14 >?-72 YES			-11 <? 66 YES	-13 <? 48 YES
10	5	89	17 -6	3	18	-1 12 2	-1 >=?-17 YES
			-1 <? 48 YES			5 <? 87 YES	-8 >?-50 YES
			-10 >?-72 YES			-17 <? 66 YES	-18 <? 48 YES
10	5	90	17 -6	3	18	0 12 3	0 >=?-17 YES
			-4 <? 48 YES			o <? 87 YES	-5 >?-50 YES
			-6 >?-72 YES			-23 <? 66 YES	-23 <? 48 YES
10	5	91	18 -6	3	19	-4 13 -1	-4 >=?-18 YES
			8 <? 51 YES			20 <? 92 YES	-17 >?-53 YES
			-22 >?-76 YES			1 <? 72 YES	-3 <? 53 YES

BIBLIOGRAPHY

[1] Phillip Griffiths and Joseph Harris.
Residues and zero-cycles on algebraic varieties.
Annals of Mathematics, 108 (1978), 461-505

[2] Joseph Harris.
A bound on the geometric genus of projective varieties.
Harvard thesis, 1977.

[3] Carlos Gómez-Mont Avalos.
Harvard thesis (in preparation).

DEPARTAMENTO DE MATEMATICAS
CINVESTAV
APARTADO POSTAL 14-740
MEXICO 07000 D.F.

Contemporary Mathematics
Volume **58**, Part I, 1986

TRANSVERSE DEFORMATIONS OF HOLOMORPHIC FOLIATIONS

XAVIER GÓMEZ-MONT

ABSTRACT. We analyze the problem of determining how many complex structures can we put on the leaves of a holomorphic foliation compatible with its transversal structure. The method we present uses the Kuranishi space of local moduli and we concentrate in computing an infinitesimal version. We show that the answer does not depend on the integrals of the foliation, but just on the cohomology of the tangent bundle to the foliation, and hence computable using methods from algebraic geometry.

A holomorphic foliation F in a complex manifold M is a decomposition of M into a disjoint union of leaves $M = \cup L_\alpha$ locally modelled on the canonical foliation of $\mathbb{C}^n$ of codimension q, $\mathbb{C}^n = \bigcup_{a \in \mathbb{C}^q} \mathbb{C}^{n-q} \times a$. Given such a holomorphic foliation F we obtain a transversely holomorphic foliation F^{tr} obtained by forgetting the complex structure of the leaves, but remembering the holomorphic manner in which the leaves are attached. In this paper we analyze the problem of how many different holomorphic structures can we put on the leaves of F^{tr} so as to produce holomorphic foliations on the compact differentiable manifold M.

Our line of approach to this problem is that each structure F and F^{tr} has a local moduli space, that is, there are two complex spaces S^{hol} and S^{tr} each parametrizing a family of foliations on M with the property that any other family of holomorphic or transversely holomorphic foliations can be obtained from these families by means of pull backs. Using these properties, we obtain a forgetful map

$$\Phi: S^{hol} \to S^{tr}$$

The main interest will be centered in calculating the derivative of this map at the point represented by F

$$(D\Phi)_F: T_F(S^{hol}) \to T_F(S^{tr})$$

We will recognize these tangent spaces from the differential geometry of F, and we will fit this map into a long exact sequence:

$$\dots \longrightarrow H^1(M,\tau) \longrightarrow T_F(S^{hol}) \xrightarrow{D\Phi} T_F(S^{tr}) \longrightarrow H^2(M,\tau) \longrightarrow \dots \qquad (1)$$

where τ is the holomorphic vector bundle of vectors tangent to the foliation F. This sequence will allow us to calculate the kernel and the cokernel of $D\Phi$, information which is relevant for the problems:

1) Given a transversely holomorphic foliation, can it be completed to a holomorphic foliation?

2) If you can, in how many different ways?

Note that the terms $H^q(M,\tau)$ in sequence (1) depend only of the holomorphic class of the tangent bundle and not in terms of the geometrical integrals L_α of the foliation. Another relevant feature is that they may be computed using methods of algebraic geometry in some particular cases.

The simplest interesting cases are foliations by analytic curves in complex surfaces. For a general such foliation F, it will have a finite number of singular points $P_1,\ldots,P_n$ all of which are locally linearizable and its analytic class completely determined by the quotient of the eigenvalues of the linear part $DX(P_i)$ of a holomorphic vector field X defining F near $P_i, \lambda_1,\ldots,\lambda_n$. If we consider a flat family F_t of foliations, $t \in T$ we obtain a Residue map

$$\mathrm{Res}: T \to \bigoplus_i \mathbb{C}$$

which assigns to F_t the local invariants at the distinct singular points $P_{i,t}$, $\mathrm{Res}(t) = (\lambda_{1,t},\ldots,\lambda_{n,t})$. We may then divide the problem of understanding the family F_t into two problems:

1) Understand the image $\mathrm{Res}(T)$. In particular, determine its codimension in $\bigoplus_i \mathbb{C}$?

2) Understand foliations with fixed analytic type at the singular points.

In the present article we restrict ourselves to the case of non-singular foliations, hence dealing exclusevely with question 2. In a forthcoming work we will analyse foliations by curves with attracting singularities, and we have thought it convinient to present a proof of the non-singular case since it illustrates our general answer for question 2, where we will obtain a similar long exact sequence (1). The transparency of the argument in the non-singular case, helps to understand the role that the singularities play in deformation theory of foliations with singularities.

1. HOLOMORPHIC AND TRANSVERSELY HOLOMORPHIC FOLIATIONS

A foliation on a differentiable manifold is defined by means of an open covering by coordinate charts $(U_\alpha,\varphi_\alpha)$ of M such that the changes of coordinates $\varphi_{\beta\alpha} = \varphi_\beta\,\varphi_\alpha^{-1}: \varphi_\alpha(U_\alpha \cap U_\beta) \to \varphi_\beta(U_\alpha \cap U_\beta)$ can be written in the form

$$\varphi_{\beta\alpha}(x_1,\ldots,x_{n-q},y_1,\ldots,y_q) = \varphi_{\beta\alpha}(\bar{x},\bar{y}) = (\varphi^1_{\beta\alpha}(x,y),\ \varphi^2_{\beta\alpha}(y)) \tag{2}$$

where φ^1 is the projection to the first $(n-q)$coordinates and φ^2 to the last q.

The condition (2) expresses the fact that the coordinate charts are leaving invariant the decomposition $y=k$. The condition (2) can also be determined by the infinitesimal condition

$$\frac{\partial \varphi^2_{\beta\alpha}}{\partial x} = 0$$

We may introduce a finer structure into the foliation by controlling more carefully the transition functions. For example we may decompose the complex vector space $\mathbb{C}^n$ as

$$\mathbb{C}^n = \bigcup_{a\in\mathbb{C}^q} \mathbb{C}^{n-q} \times \{a\}$$

If we want to introduce a holomorphic foliation into a manifold by charts we require the $\varphi_{\beta\alpha}$ to be holomorphic in its variables and besides, to satisfy condition (2) where $(\bar{x},\bar{y}) \in \mathbb{C}^{n-q}\times\mathbb{C}^q$.

We could also consider the decomposition

$$\mathbb{R}^m\times\mathbb{C}^q = \bigcup_{a\in\mathbb{C}^q} \mathbb{R}^m\times\{a\}$$

To introduce this structure on M we require condition (2) where $(\bar{x},\bar{y}) \in \mathbb{R}^m \times \mathbb{R}^{2q}$ and besides the function φ^2 should be a holomorphic function as a function from $\mathbb{R}^{2q}$ to $\mathbb{R}^{2q}$. We will call such a structure a transversely holomorphic foliation. For more details, see [1] or [2].

2. INFINITESIMAL AUTOMORPHISMS.

Let F be a foliation in the manifold M. We want to consider those vector fields defined on open subsets of M such that the local 1-parameter group that they generate by integration preserves F; i.e. sends leaves to leaves and if we are considering a G-foliation, then it should also preserve this additional G-structure, where G is holomorphic or transversely holomorphic.

Let us analize the local model first. Consider in $\mathbb{R}^n$ coordinates $(x,y) = (x_1,\ldots,x_{n-q},y_1,\ldots,y_q) \in \mathbb{R}^{n-q}\times\mathbb{R}^q$ and the foliation defined by y=constant. A vector field X on an open set of U may be written as

$$X = \sum_{i=1}^{n-q} f_i(x,y)\frac{\partial}{\partial x_i} + \sum_{i=n-q+1}^{n} f_i(x,y)\frac{\partial}{\partial y_i} \tag{3}$$

Let φ_t be the local 1-parameter group generated by X. Fix t_0 and y_0, and view

$$x \to \varphi_{t_0}(x,y_0)$$

a function $\mathbb{R}^{n-q}$ to $\mathbb{R}^n$, which tells us where in $\mathbb{R}^n$ the leaf $\mathbb{R}^{n-q}\times y_0$ lies after flowing along X a time t_0.

If we want that the leaf be mapped to another leaf then the last coordinates $\varphi^2_{t_0}(x,y_0)$ should be a constant, so

$$\frac{d}{dx}\varphi^2_{t_0}(x,y_0) = 0$$

So if the flow preserves the foliation, this identity is true for every x,y and t. So we also have

$$\frac{d}{dt}\frac{d}{dx}\varphi^2 = 0$$

If X is of class C^2, this is equivalent to

$$0 = \frac{d}{dx}\frac{d}{dt}\varphi^2 = \frac{d}{dx}\sum_{i=n-q+1}^{n} f_i\frac{\partial}{\partial y_i}$$

so we obtain the necessary conditions

$$\frac{\partial f_i}{\partial x_j} = 0 \qquad \begin{array}{l} i = n-q+1,\dots,n \\ j = 1,\dots,n-q \end{array}$$

The converse is also true as can be checked by integration. We have proved:

LEMMA 1: *A local vector field* X *preserves the foliation on* $\mathbb{R}^n$ *given by* y=*constant if and only if it can be written as*

$$X = \Sigma f_i(x,y)\frac{\partial}{\partial x_i} + \Sigma f_i(y)\frac{\partial}{\partial y_i} \tag{4}$$ □

Now we want to give an intrinsic characterization of the infinitesimal automorphisms as in (4) that will lift to manifolds.

If F is a foliation on M, then every point p of M lies on a leaf L_p, and the tangent space to this leaf is an $(n-q)$-dimensional vector subspace $T_p(F) \subset T_p(M)$. With the help of the local coordinates we see that $T(F)$ forms a vector sub-bundle of the tangent bundle of M called the tangent bundle to the leaves. Define the normal bundle as the quotient bundle on M

$$0 \to T(F) \to T(M) \to N(F) \to 0$$

It is a q-dimensional vector bundle.

In local trivializing coordinates (x,y), $T(F)$ is generated by $\frac{\partial}{\partial x_1},\dots,\frac{\partial}{\partial x_{n-q}}$ and $N(F)$ is obtained by projecting the vector field (3) to the components $\frac{\partial}{\partial y_i}$. Hence, the infinitesimal automorphisms are characterized as those vector fields such that when projected to the normal bundle are constant along the leaves. A closer analysis of the meaning of the infinitesimal automorphisms of F is best understood in the language of sheaves.

Let F be a holomorphic foliation of the compact complex manifold M, let Θ_M^{hol} be the sheaf of holomorphic vector fields on M, T_F^{hol} the sheaf of holomorphic tangent vector fields to the foliation; the normal sheaf N_F is defined by the exact sequence

$$0 \to T_F^{hol} \to \Theta_M^{hol} \to N_F \to 0$$

Let Θ_F^{hol} be the sheaf of holomorphic infinitesimal automorphisms of F, the sheaf of sections of the normal bundle constant along the leaves is denoted by Θ_F^{tr} and defined by the exact sequence

$$0 \to T_F^{hol} \to \Theta_F^{hol} \to \Theta_F^{tr} \to 0 \tag{5}$$

Now let G be a transversely holomorphic foliation on M, Θ_M^{∞} the sheaf of C^{∞} vector fields, T_G^{∞} the sheaf of tangent vectors to the foliation and $\widetilde{\Theta}_G^{tr}$ the sheaf of infinitesimal automorphisms of G. Denote by Θ_G^{tr} those sections of the normal bundle to G that are constant along the leaves and holomorphic in the transversal variables. Projecting to the normal bundle we obtain a similar sequence to (5)

$$0 \to T_G^{\infty} \to \widetilde{\Theta}_G^{tr} \to \Theta_G^{tr} \to 0 \tag{6}$$

THEOREM 2: *Let* F *be a holomorphic foliation on* M, *and* F^{tr} *the transversely holomorphic foliation associated to* F *by forgetting the complex structure of the leaves. The inclusion* $i:(M,F) \to (M,F^{tr})$ *induces a commutative diagram of sheaves on* M:

$$\begin{array}{ccccccccc} 0 & \longrightarrow & T_F^{hol} & \longrightarrow & \Theta_F^{hol} & \longrightarrow & \Theta_F^{tr} & \longrightarrow & 0 \\ & & \downarrow i_1 & & \downarrow i_2 & & \downarrow i_3 & & \\ 0 & \longrightarrow & T_{F^{tr}}^{\infty} & \longrightarrow & \widetilde{\Theta}_{F^{tr}}^{tr} & \longrightarrow & \Theta_{F^{tr}}^{tr} & \longrightarrow & 0 \end{array} \tag{7}$$

and 1) i_3 is an isomorphism

2) The following long sequence is exact

$$\begin{array}{l} 0 \longrightarrow H^0(M,T_F^{hol}) \longrightarrow H^0(M,\Theta_F^{hol}) \longrightarrow H^0(M,\Theta_F^{tr}) \longrightarrow \\ \longrightarrow H^1(M,T_F^{hol}) \longrightarrow H^1(M,\Theta_F^{hol}) \xrightarrow{i_*} H^1(M,\widetilde{\Theta}_{Ftr}^{tr}) \longrightarrow \\ \cdots \longrightarrow H^k(M,T_F^{hol}) \longrightarrow H^k(M,\Theta_F^{hol}) \xrightarrow{i_*} H^k(M,\widetilde{\Theta}_{Ftr}^{tr}) \longrightarrow \cdots \end{array} \tag{8}$$

where i_* are the maps induced in cohomology by i_2.

3) The cohomology groups $H^q(M,\Theta_F^{hol})$, $H^q(M,\Theta_F^{tr})$ for $q \geqslant 0$ and $H^q(M,\widetilde{\Theta}_{Ftr}^{tr})$ for $q > 0$ are finite dimensional vector spaces.

Proof: The map i is defined in local charts by $i: \mathbb{C}^{n-q} \times \mathbb{C}^q \to \mathbb{R}^{2(n-q)} \times \mathbb{C}^q$, $i(x,y) = (x,y)$ where x is thought of as a complex variable in $\mathbb{C}^{n-q}$ in the domain, and a real variable in $\mathbb{R}^{2(n-q)}$ in the codomain. By Lemma 1 the elements of Θ_F^{hol} and $\tilde{\Theta}_{Ftr}^{tr}$ may be written as

$$\Sigma f_i(x,y)\frac{\partial}{\partial x_i} + \Sigma g_i(y)\frac{\partial}{\partial y_i}$$

where f_i and g_i are holomorphic for Θ_F^{hol} and f_i is C^∞ and g_i holomorphic for Θ_{Ftr}^{tr}. The elements of Θ_F^{tr} and Θ_{Ftr}^{tr} consists only of the g_i's that are holomorphic for both. This proves that i_3 is an isomorphism.

Since τ_{Ftr}^∞ is a fine sheaf, we obtain for $q > 0$ the isomorphisms

$$H^q(M,\tilde{\Theta}_{Ftr}^{tr}) \simeq H^q(M,\Theta_{Ftr}^{tr}) \simeq H^q(M,\Theta_F^{tr})$$

Incorporating these isomorphisms into the long exact sequence of the top of (7) gives (8).

A proof that $H^q(M,\Theta_F^{tr}) = H^q(M,\Theta_{Ftr}^{tr})$ is finite dimensional may be found in [1] or [3].

$H^q(M,\tau)$ is finite dimensional since M is compact and τ coherent. The long exact sequence in (7) gives then that $H^q(M,\Theta_F^{hol})$ is finite dimensional for $q \geqslant 0$ and using again (7) and the fact that τ_F^∞ is a finite sheaf we conclude that $H^q(M,\tilde{\Theta}_F^{hol})$ is finite dimensional for $q > 0$. □

REMARK: Deformation theory will tell us to look carefully at the map $i_{2*}: H^1(M,\Theta_F^{hol}) \to H^1(M,\tilde{\Theta}_{Ftr}^{tr})$ obtained from the inclusion of the infinitesimal automorphisms of each structure. Sequence (7) is then saying that since τ_{Ftr}^∞ is fine and i_3 is an isomorphism, we may compute it from the cohomology of the top row, but then the Kernel and coKernel may be estimated using the cohomology of the coherent sheaf τ:

$$\cdots \longrightarrow H^1(M,\tau) \longrightarrow H^1(M,\Theta_F^{hol}) \xrightarrow{i_{2*}} H^1(M,\tilde{\Theta}_{Ftr}^{tr}) \longrightarrow H^2(M,\tau) \longrightarrow \cdots \tag{9}$$

The interpretation of i_{2*} in terms of deformation theory, together with the computability of the cohomology of τ, is what makes the sequence (9) fruitful.

3. DEFORMATION THEORY FOR FOLIATIONS

A *deformation of the foliation* F in the compact manifold M is given by the following data:

1) A differentiable fibre bundle $\pi: D \to S$ with fiber diffeomorphic to M. D is the deformation space and S is the parameter space of the deformation.

2) A coordinate cover $\{U_i\}$ of D with coordinate charts $\varphi_i: U_i \to \mathbb{R}^p \times \mathbb{R}^{n-q} \times \mathbb{R}^q$, $\varphi_i(p) = (x, y_1, y_2)$ such that the transition functions have the form

$$\varphi_{ji}(x,y_1,y_2) = (\varphi^1_{ji}(x),\ \varphi^2_{ji}(x,y_1,y_2),\ \varphi^3_{ji}(x,y_2))$$

The projection into the first factor defines local coordinates for the bundle structure π, and the other conditions mean that for each x fixed we obtain a foliation on $\pi^{-1}(x)$, which we denote by F_x.

3) $F = F_0$ for some point $0 \in S$.

If F is to be *holomorphic or transversely holomorphic*, then we require that the transition functions φ^2 and φ^3 be holomorphic in (y_1, y_2) or φ^3 in y_2 respectively. This means that each foliation F_t is holomorphic or transversely holomorphic.

In any one of this two cases we will say that the *deformation is holomorphic* if

a) For the holomorphic case: x, y_1, y_2 are complex variables and all transition functions are holomorphic in all its variables.

b) For the transversely holomorphic case: x,y are complex variables and φ^1 and φ^3 are holomorphic on its variables. For holomorphic foliations, the parameter spaces may be complex analytic spaces, and the deformation are flat.

We are interested in analyzing all deformations parametrized by neighborhoods of 0 with $F_0 = F$. When we are interested only in the deformation restricted to a neighborhood of 0, arbitrarily small, we then call it a *germ of a deformation.* Two deformations $(D,S.0)$ and $(D',S',0')$ of F are isomorphic if there exists a diffeomorphism $\pi\colon D \to D'$ preserving the fibre bundle structure

$$\begin{array}{ccc} D & \xrightarrow{\varphi} & D' \\ \pi\downarrow & & \downarrow\pi \\ S & \xrightarrow{\bar\varphi} & S' \end{array}$$

such that for every point $s \in S$, the restriction of F to an arbitrary fiber

$$\varphi\colon \pi^{-1}(s) \to \pi^{-1}(\bar\varphi(s))$$

is an automorphism of the foliated manifolds; i.e. F_s and $F_{\bar\varphi(s)}$ are isomorphic as foliations by φ.

In case that the foliations are holomorphic or the deformations are holomorphic then the corresponding maps should be holomorphic.

Is one natural construction technique to obtain deformations from known deformations. Let $\pi\colon D \to S$ be a deformation and $f\colon S' \to S$ be a smooth map. From the theory of fiber bundles we may construct over S' a bundle $D' = f^*(D)$ and maps

$$\begin{array}{ccc} D' & \xrightarrow{f'} & D \\ \pi'\downarrow & & \downarrow\pi \\ S' & \xrightarrow{f} & S \end{array}$$

$$D' = \{(s,p) \in S' \times D / f(s) = \pi(p)\} \subset S' \times D$$

where the first and second projections are the maps π' and f'.

With the coordinate charts on D its is very simple to give coordinate charts on D' so that it becomes a deformation. Of course, now we have that the foliations F'_{s_1} and F'_{s_2} in D' will be isomorphic if $f(s_1) = f(s_2)$, since they are both isomorphic to $F_{f(s_1)}$.

Let $\pi: D \to S$ be a deformation of a holomorphic foliation F in the compact manifold M. Let Θ_D denote the sheaf of vector fields on D such that the local 1-parameter group that it generates sends fibers of π into fibers of π, and when restricted to one of the above fibers, is an automorphism of the underlying holomorphic foliation. If (x,y_1,y_2) are local coordinates where x is the parameter at s and the foliation on the fiber $x = x_0$ is given by y_2 = constant, then these vector fields have the form

$$X = f_1(x)\frac{\partial}{\partial x} + f_2(x,y_1,y_2)\frac{\partial}{\partial y_1} + f_3(x,y_2)\frac{\partial}{\partial y_2}$$

with f_2 and f_3 holomorphic functions of y_1 and y_2 for every fixed value of x.

Let $\pi^{-1}(\Theta_S)$ be the sheaf on D obtained by pulling back to D the sheaf of tangent vectors to S. A section Y of $\pi^{-1}(\Theta_S)$ on an open set U gives, for every point $p \in U$, a vector at $\pi(p) \in S$ in such a way that it is the same vector if p and q lie on the same fiber, i.e.

$$Y(p) = Y(q) \quad \text{if} \quad \pi(p) = \pi(q)$$

In the local expresion above, it is expressed as

$$f_1(x)\ \frac{\partial}{\partial x}$$

Hence we have a natural sheaf mapping

$$\Theta_D \to \pi^{-1}(\Theta_S)$$

which arise by projecting into the tangent space of S. We may complete to an exact sequence of sheaves

$$0 \to \Theta_{D/S} \to \Theta_D \to \Theta^{-1}(\Theta_S) \to 0, \tag{10}$$

where in the kernel we have the sheaf of infinitesimal automorphisms of each fiber, in local coordinates

$$f_2(x,y_1,y_2)\frac{\partial}{\partial y_1} + f_3(x,y_2)\frac{\partial}{\partial y_3}$$

We may also consider the families of sequences (5) to give

$$\begin{array}{ccccccccc} & & 0 & & & & & & \\ & & \downarrow & & & & & & \\ & & \tau^{hol}_{D/S} & & & & & & \\ & & \downarrow & & & & & & \\ 0 & \longrightarrow & \Theta^{hol}_{D/S} & \longrightarrow & \Theta^{hol}_{D} & \longrightarrow & \pi^{-1}(\Theta_S) & \longrightarrow & 0 \\ & & \downarrow & & & & & & \\ & & \Theta^{tr}_{D/S} & & & & & & \\ & & \downarrow & & & & & & \\ & & 0 & & & & & & \end{array} \tag{11}$$

In case $\pi': D' \to S'$ is a deformation of transversely holomorphic foliations, we obtain from (6) and (10) the similar sequence

$$\begin{array}{ccccccccc} & & 0 & & & & & & \\ & & \downarrow & & & & & & \\ & & \tau'^{\infty}_{D'/S'} & & & & & & \\ & & \downarrow & & & & & & \\ 0 & \longrightarrow & \widetilde{\Theta}_{D'/S'} & \longrightarrow & \widetilde{\Theta}_{D'} & \longrightarrow & \pi'^{-1}(\Theta_{S'}) & \longrightarrow & 0 \\ & & \downarrow & & & & & & \\ & & \Theta^{tr}_{D'/S'} & & & & & & \\ & & \downarrow & & & & & & \\ & & 0 & & & & & & \end{array} \tag{12}$$

The basic obstruction that one measures in deformation theory is to lifting sections of $\pi^{-1}(\Theta_S)$ in (11) and (12). Formally, one considers the exact sequence of direct image sheaves of the rows to give

$$\begin{array}{l} 0 \to \pi_*\Theta^{hol}_{D/S} \to \pi_*\Theta^{hol}_{D} \to \Theta_S \xrightarrow{\rho} R^1\pi_*\Theta^{hol}_{D/S} \to \cdots \\ 0 \to \pi_*\widetilde{\Theta}^{hol}_{D'/S'} \to \pi_*\widetilde{\Theta}^{tr}_{D'} \to \Theta_S \xrightarrow{\rho'} R^1\pi_*\widetilde{\Theta}^{tr}_{D'/S'} \to \cdots \end{array} \tag{13}$$

where ρ and ρ' are the *Kodaira-Spencer map* of the family (see [4] p. 360 for more details). Evaluating at some point s_0 we obtain the *map of infinitesimal deformations*

$$\rho: T_{s_0}S \longrightarrow H^1(M,\Theta^{hol}_F)$$

$$\rho': T_{s_0}S' \to H^1(M,\widetilde{\Theta}^{tr}_F)$$

A *versal deformation* of a foliation F on the compact manifold M is deformation $\pi: D \to S$ with $F_0 = F$ such that for any other deformation $\pi': D' \to S'$ with $F_0 = F$ there exists a map $f: S' \to S$ such that D' is isomorphic to $f^*(D)$ as germs of deformations on $(S',0)$ and the tangent map

$$df: T_{0'}S' \to T_0S$$

is unique. The main point of the theory of deformations is the construction of a versal deformation and analyzing its properties. We will rely heavily on the following existence

THEOREM A([2],[5]): *Let* F *be a transversely holomorphic (or holomorphic) foliation on a compact manifold* M, *then there is a germ of an analytic space* $(S,0)$ *parametrizing a germ of a deformation* F_S *such that for any germ* $F_{S'}$ *of holomorphic deformations of* F *parametrized by* $(S',0)$, *there is a holomorphic map* $\varphi: (S',0) \to (S,0)$ *such that the induced deformation* $F_{\varphi(S')}$ *is isomorphic to* $F_{S'}$. *Moreover the differential* $d_0\varphi$ *of* φ *at* 0 *is unique, and the Kodaira-Spencer map* $\rho: T_0S \to H^1(M,\widetilde{\Theta}_F^{tr})$ *(respectively* $\rho: T_0S \to H^1(M,\Theta_F^{hol})$) *is an isomorphism.*

4. TRANSVERSE DEFORMATIONS OF HOLOMORPHIC FOLIATIONS

Let $\pi: D \to S$ be a holomorphic deformation of the holomorphic foliation F on M, and denote by $\pi^{tr}: D^{tr} \to S$ the family of transversely holomorphic foliations obtained by forgetting the complex structure of the leaves, as in Theorem 2.

From sequence (11), (12) and (13) we obtain

$$\begin{array}{ccc} & & R^1\pi_*\tau^{hol}_{D/S} \\ & & \downarrow \\ \Theta_S & \xrightarrow{\rho} & R^1\pi_*\Theta^{hol}_{D/S} \\ & & \downarrow \\ & & R^1\pi_*\Theta^{tr}_{D/S} \\ & & \downarrow \\ & & R^2\pi_*\tau^{hol}_{D/S} \end{array}$$

and

$$\begin{array}{ccc} \Theta_S & \xrightarrow{\rho^{tr}} & R^1\pi_*\widetilde{\Theta}^{tr}_{D^{tr}/S} \\ & & \cong \downarrow \alpha \\ & & R^1\pi_*\Theta^{tr}_{D^{tr}/S} \end{array}$$

where α is an isomorphism since τ^∞ is a fine sheaf.

By theorem 2 we have an isomorphism $\Theta^{tr}_{D/S} \cong \Theta^{tr}_{D^{tr}/S}$ so we obtain a commutative diagram:

$$\begin{array}{ccl} & & R^1\pi_*\tau^{hol}_{D/S} \\ & & \downarrow \\ \Theta_S & \xrightarrow{\ \rho\ } & R^1\pi_*\Theta^{hol}_{D/S} \\ & \searrow^{\rho^{tr}} & \downarrow \\ & & R^1\pi_*\tilde{\Theta}^{tr}_{D^{tr}/S} \\ & & \downarrow \\ & & R^2\pi_*\tau^{hol}_{D/S} \end{array} \tag{14}$$

where ρ and ρ^{tr} are the Kodaira-Spencer maps of the family.

Denote by

$$\pi\colon D \longrightarrow S^{hol} = S \qquad \pi'\colon D' \longrightarrow S^{tr} = S'$$

The two versal families associated to the holomorphic foliation F by theorem A, viewed as families of holomorphic and transversely holomorphic foliations. From the versal properties of S^{tr}, and viewing the family π as a family of transversely holomorphic foliations, we obtain a holomorphic map f such that

$$\begin{array}{ccccc} D^{tr} & \xrightarrow{\ \tilde f\ } & f^*(D') & & D' \\ & {\scriptstyle \pi=\pi^{tr}}\searrow & \swarrow & & \downarrow{\scriptstyle \pi'} \\ & & S & \xrightarrow{\quad f \quad} & S' \end{array}$$

where $\tilde f$ is a transversely holomorphic isomorphism, that identifies D^{tr} and $f^*(D')$. The derivative of f and (13) give

$$\begin{array}{ccccc} \Theta_S & \xrightarrow{\ \rho^{tr}\ } & R^1\pi_*\tilde{\Theta}^{tr}_{D^{tr}/S} & & \\ {\scriptstyle df}\downarrow & & \cong\downarrow{\scriptstyle f} & & \\ f^*\Theta_S & \longrightarrow & f^*R^1\pi'_*\tilde{\Theta}^{tr}_{D'/S'} & = & R^1\pi_*\tilde{\Theta}^{tr}_{D^{tr}/S} \end{array} \tag{15}$$

substituting (15) in (14) and evaluating at $t = 0$ we obtain by flatness

$$\begin{array}{ccl} & & H^1(M,\tau^{hol}_F) \\ & & \downarrow \\ T_0S & \xrightarrow{\ \rho_0\ } & H^1(M,\Theta^{hol}_F) \\ {\scriptstyle df_0}\downarrow & & \downarrow \\ T_0S' & \xrightarrow{\ \rho'_0\ } & H^1(M,\tilde{\Theta}^{tr}_F) \\ & & \downarrow \\ & & H^2(M,\tau^{hol}_F) \end{array}$$

where the two horizontal maps are isomorphisms by the versality of the families and the vertical sequence is the sequence in (9). We have proved.

THEOREM 3: *Let* F *be a holomorphic foliation in the compact manifold* M, S^{hol} *and* S^{tr} *the versal families of deformation,* $\Phi\colon S^{hol} \to S^{tr}$ *the forgetful map.*

Then the exact sequence (9) gives rise to the exact sequence

$$\cdots \longrightarrow H^1(M,\tau^{hol}) \longrightarrow T_0(S^{hol}) \xrightarrow{\Phi} T_0(S^{tr}) \longrightarrow H^2(M,\tau^{hol}) \longrightarrow \cdots \qquad (1)$$

COROLLARY 4: *If* $H^1(M,\tau) = 0$ *and* S *is locally irreducible at* 0 *then* Φ *is injective as a germ of a map.*

5. RULED SURFACES.

We exemplify theorem 3 in foliations of ruled surfaces.

Let $\pi_1(S)$ be the fundamental group of the compact orientable surface S and

$$\rho : \pi_1(S) \to PSL(2,\mathbb{C})$$

a representation of $\pi_1(S)$ into the group of conformal automorphisms of the Riemann sphere $\bar{\mathbb{C}}$. If we suspend the representation

$$M = S\times_\rho \bar{\mathbb{C}}$$

where $\tilde{S}$ is the universal cover of S and the suspension construction is the quotient of $\tilde{S}\times\bar{\mathbb{C}}$ by the diagonal action of $\pi_1(S)$. The transversely holomorphic foliation

$$\bigcup_{z\in\bar{\mathbb{C}}} \tilde{S}\times\{z\} = \tilde{S}\times\bar{\mathbb{C}}$$

descends to M to give a transversely holomorphic foliation F^{tr}. Its leaves are transverse to the ruling.

Choose also a conformal structure in S so that M is a complex surface and the foliation F^{hol} is holomorphic.

We have a map $\rho: M \to S$ which expresses M as a holomorphic ruled surface over S, the tangent bundle to the leaves τ is clearly $\rho^*(\Theta_S)$ where Θ_S is the tangent sheaf of S. We may then compute the cohomology of τ with the help of the spectral sequence associated to ρ. A simple calculation gives $H^q(M,\tau) = H^q(S,\Theta_S)$, hence is $3g - 3$ if $q = 1$ and 0 if $q = 2$, so (1) becomes

$$H^1(S,\Theta_S) \to H^1(M,\Theta_F^{hol}) \to H^1(M,\Theta_F^{tr}) \to 0$$

With the help of [2] it has the following interpretation: $H^1(M,\Theta_F^{tr})$ represent the deformation of the representation and $H^1(S,\Theta_S)$ is the deformation of the conformal structure of S. Actually we have a decomposition

$$H^1(M,\Theta_F^{hol}) = H^1(S,\Theta_S) \oplus H^1(M,\Theta_F^{tr})$$

since we may vary completely independently the conformal structure on S and the representation, the map

$$\Phi: S^{hol} \to S^{tr}$$

is locally at 0 a product bundle with fiber a ball of dimension $3g-3$ if the genus of S is greater than one.

BIBLIOGRAPHY

[1] Duchamp, T., Kalka, M. *Deformation theory for holomorphic foliations.* J. Diff. Geometry, 14(1979), 317-337.

[2] Girbau, J., Haefliger, A., Sundararaman, D. *On deformations of transversely holomorphic foliations.* Crelle Journal, (345), 1983, 122-147.

[3] Gómez-Mont, X. *Transversal Holomorphic Structures.* J. Diff. Geometry, 15(1980), 161-186.

[4] Kodaira, L., Spencer, D. *On deformations of complex analytic structures.* I, II Ann. Math. 67(1958), 328-466.

[5] Kuranishi, M. *Deformations of compact complex manifolds.* Université de Montreal, Canadá, 1971.

INSTITUTO DE MATEMATICAS
Universidad Nacional Autónoma de México
México, 04510, D.F.
MEXICO

Contemporary Mathematics
Volume 58, Part I, 1986

SINGULARITIES OF ALGEBRAIC SURFACES AND CHARACTERISTIC NUMBERS

F. Hirzebruch

ABSTRACT. The Chern numbers c_1^2 and c_2 of an algebraic surface of general type satisfy the Miyaoka-Yau inequality $c_1^2 \leqq 3c_2$. If the surface contains rational or elliptic curves, then $3c_2 - c_1^2$ is positive and one can give an estimate from below using the rational and elliptic curves. This has many geometric applications. I report on work of R. Kobayashi and Y. Miyaoka and use the Bonn dissertation of Th. Höfer and the Bonn Diplomarbeit of K. Ivinskis.

1. For every compact smooth algebraic surface X the Chern numbers c_2 and c_1^2 are defined. The Chern number c_2 equals the Euler number e of X, whereas c_1^2 is the selfintersection number of a canonical divisor of X. If X is a surface of general type, then $c_2 > 0$ and

$$c_1^2 \leqq 3c_2 \quad .$$

The famous inequality $c_1^2 \leqq 3c_2$ has a long history. It was proved by Y. Miyaoka [15] using ideas due to F.A. Bogomolov (compare A. Van de Ven [20]). It was proved independently by differential-geometric methods for the case that the canonical bundle of X is ample: According to T. Aubin and S.T. Yau there exists a unique Einstein-Kähler metric on X and by a result of H. Guggenheimer (1952) the difference $3c_2 - c_1^2$ is then given in terms of this metric by an integral over X with non-negative integrand which measures the deviation from constant holomorphic sectional curvature. Therefore $c_1^2 \leqq 3c_2$ (Yau [22]). Furthermore (Yau [22]) the equation $c_1^2 = 3c_2$ implies that the universal cover of X is the ball

$B = \{z \in \mathbb{C}^2 : |z_1|^2 + |z_2|^2 < 1\}$. This important conclusion was proved by Yau and Miyaoka also in the case that the canonical bundle is not necessarily ample (see [16]).

We refer to the Bourbaki lecture of J.-P. Bourguignon [1] for further references to the literature.

2. As explained above, $c_1^2 \leqq 3c_2$ for a surface X of general type, and $c_1^2 = 3c_2$ if and only if the universal cover of X is the ball B . (The "if" part is the "proportionality" of [7].) If X contains rational or elliptic curves, then, clearly, the universal cover of X cannot be the ball and hence $c_1^2 < 3c_2$. Therefore, to a configuration E of rational or elliptic curves a positive number $m(E)$ should be assigned such that

$$3c_2 - c_1^2 \geqq m(E) \quad . \tag{1}$$

If E is a disjoint union of finitely many smooth elliptic curves C_j on X , then according to F. Sakai [19]

$$3c_2 - c_1^2 \geqq \sum(-C_j^2) \quad . \tag{2}$$

The selfintersection numbers C_j^2 are negative, since X is a surface of general type. Y. Miyaoka [17] has studied numbers $m(E)$ for configurations of rational curves. I discussed such questions with him during one of his visits in Bonn.

Let us consider an example. Suppose $S_1, \dots, S_r$ is a chain of smooth rational curves, where S_j intersects S_{j+1} transversally in exactly one point $(j=1,\dots,r-1)$ and $S_iS_j = 0$ for $|j-i| > 1$. We assume $-S_jS_j = b_j \geqq 2$. The dual graph of this chain is

(3)

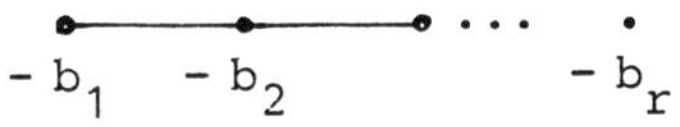

For the continued fractions

$$b_1 - \frac{1|}{|b_2} - \frac{1|}{|b_3} - \cdots - \frac{1|}{|b_r} = \frac{n}{q}$$

$$b_r - \frac{1|}{|b_{r-1}} - \frac{1|}{|b_{r-2}} - \cdots - \frac{1|}{|b_1} = \frac{n}{q'}$$

we have

$$0 < q < n \quad , \quad 0 < q' < n \quad , \quad qq' \equiv 1 (\mathrm{mod}\, n) \quad .$$

In fact, this chain of rational curves is the resolution of the quotient singularity $\mathbb{C}^2/\mu_n$ where the group

$$\mu_n = \{\alpha \in \mathbb{C} : \alpha^n = 1\}$$

acts by $\alpha(z_1,z_2) = (\alpha z_1, \alpha^q z_2)$. Compare [6]. For the chain (3) of rational curves we can define the local Euler number e_{loc} which equals $r+1$. It is the Euler number of r twodimensional spheres $S_1,\dots,S_r$ where a point on S_j is identified with a point of S_{j+1} for $j = 1,\dots, r-1$. We also have a local canonical divisor $K_{loc} = \sum_{i=1}^{r} c_i S_i$ where the rational numbers c_i are defined by the linear equations

$$K_{loc}\, S_i + S_i S_i = -2$$

$$\sum_{i=1}^{r} c_i S_i S_j = b_j - 2$$

corresponding to the adjunction formula.

The intersection matrix $S_i S_j$ is negative-definite of determinant $(-1)^r n$.

Let E be the chain (3) . We define

$$m(E) = 3\, e_{loc} - K_{loc}^2 - \frac{3}{n}$$

and expect an inequality (1). Why? Let X' be the surface obtained by collapsing E to a point. It has one singular point.

Suppose (as a "Gedankenexperiment") that there is a smooth surface Y with an action of μ_n which is free outside one point p such that $Y/\mu_n = X'$. Then

$$3c_2(Y) - c_1^2(Y) = n(3c_2(X) - c_1^2(X) - m(E))$$

and (1) follows for X because Y is smooth of general type and satisfies the Miyaoka-Yau inequality. This is no proof at all, but motivates the definition of $m(E)$ and indicates how to look for a proof.

In his recent Bonn Diplomarbeit K. Ivinskis [11] has given a nice formula for $m(E)$ for a chain (3). We have

$$(4) \qquad m(E) = 1 + \sum_{i=1}^{r} (b_i+1) + \frac{q+q'-1}{n} .$$

Similar formulas can be written down for all quotient singularities, i.e. for all configurations of rational curves arising from the resolution of quotient singularities (compare [2]). This becomes especially simple for the rational double points (see 8. below), i.e. for the configurations A_r, D_r, E_6, E_7, E_8. Here all curves have selfintersection number -2 which implies $K_{loc} = 0$ and $m(E) = 3e_{loc} - \frac{3}{|G|}$ where G is the finite group acting on $\mathbb{C}^2$ (and freely on $\mathbb{C}^2 - \{0\}$) such that $\mathbb{C}^2/G$ gives the quotient singularity. For example, A_r is the configuration (3) with $b_i = 2$ and

$$(5) \qquad m(A_r) = 3(r+1) - \frac{3}{r+1}$$

which agrees with (4) since $n = r+1$, $q = q' = r$. For E_8 we have $e_{loc} = 9$ and

$$m(E_8) = 27 - \frac{3}{120} = 27 - \frac{1}{40} .$$

3. THEOREM. Let X be a smooth surface of general type and $E_1,\ldots,E_k$ configurations (disjoint to each other) of rational curves (arising from quotient singularities) and $C_1,\ldots,C_p$ smooth elliptic curves (disjoint to each other and disjoint to the E_i). Let c_1^2, c_2 be the Chern numbers of X. Then

$$(6) \qquad 3c_2 - c_1^2 \geq \sum_{i=1}^{k} m(E_i) + \sum_{j=1}^{p} (-c_j^2) .$$

This is a part of a theorem of Miyaoka [17]. It includes the older result of F. Sakai on elliptic curves. The inequality (6) is already true if the Kodaira dimension of X is non-negative.

4. R. Kobayashi ([13],[14]) has developed a theory of Einstein-Kähler metrics for V-manifolds (surfaces with quotient singularities) which are not necessarily complete (they are surfaces with elliptic curves removed). S.T. Yau told me that he also studied

these problems. R. Kobayashi was able to prove the inequality (6) and to prove a theorem for the case that equality holds in (6). He had to introduce some additional assumptions concerning (-1)-curves and (-2)-curves, i.e., smooth rational curves with selfintersection number -1 or -2 respectively. We simplify by making these assumptions unnecessarily strong.

THEOREM. Let X, E_i, C_j be as in the above theorem. Suppose that X is minimal, i.e. does not contain (-1)-curves and that each (-2)-curve is contained in one configuration E_i. If equality holds in (6), then there exists a discrete subgroup Γ of the group of automorphisms of the ball, such that $X' - \cup C_j = \Gamma\backslash B$. Here X' is the singular surface obtained from X by blowing down the E_i. The group Γ has only isolated points in B with non-trivial isotropy group. They give the quotient singularities of X'. The group Γ has p "cusps". If one compactifies $\Gamma\backslash B$ to $\overline{\Gamma\backslash B}$ by adding p points at infinity for the p cusps, one gets p singular points which are resolved by the p elliptic curves C_j. Thus X is the smooth model (minimal desingularization) of $\overline{\Gamma\backslash B}$.

J.C. Hemperly [5] was the first to study the singularities at the cusps of surfaces $\overline{\Gamma\backslash B}$. An extensive study of the surfaces $\overline{\Gamma\backslash B}$ if Γ is a Picard modular group was carried out in many papers by R.-P. Holzapfel.

5. AN EXAMPLE. Consider the following surface Y in $P_4(\mathbb{C})$ with homogeneous coordinates $x_0, x_1, \ldots, x_4$.

$$\sum_{i=0}^{4} x_i^5 = 0 \quad , \quad \sum_{i=0}^{4} x_i^{15} = 0 \quad .$$

It covers the cubic surface S

$$\sum_{i=0}^{4} u_i = 0 \quad , \quad \sum_{i=0}^{4} u_i^3 = 0$$

which is the Clebsch diagonal surface (Clebsch 1871). It was studied in [8] in relation to Hilbert modular surfaces. The covering map $Y \longrightarrow S$ is given by $u_i = x_i^5$ and has degree 5^4. The cubic surface S is smooth. The hyperplane section $u_i = 0$ is a cubic curve consisting of 3 lines intersecting each other in 3 points. This determines 15 distinguished points of S (the point $0:1:-1:1:-1$

and all permutations) over which we have $15 \cdot 5^3 = 1875$ points of Y which are singular, in fact they are A_4-singularities. The three hyperplane sections $u_1 = 0$, $u_2 = 0$, $u_3 = 0$ intersect in exactly one point. We get 10 distinguished points ((0:0:0:1:-1) and permutations). They are the 10 Eckardt points (points of intersection of three lines on the surface). Over each Eckardt point we have 5 points of Y. They are singular. Each of these 50 singularities is a cone over a Fermat curve of degree 5 (genus 6). We now pass to the smooth model X of Y by resolving all singularities in the minimal way and want to calculate the Chern numbers of X. For a complete intersection X_0 of two hypersurfaces of degrees 5 and 15 in $P_4(\mathbb{C})$ in general position the total Chern class is given by

$$1 + c_1(X_0) + c_2(X_0) = (1+5g+10g^2)(1+5g)^{-1}(1+15g)^{-1}$$

where $g \in H^2(X_0,\mathbf{Z})$ is the cohomology class of a hyperplane section. This gives

$$1 + c_1(X_0) + c_2(X_0) = (1+10g^2)(1+15g)^{-1}$$

$$c_1(X_0) = -15g \ , \ c_2(X_0) = 235g^2$$

$$c_1^2(X_0) = 75 \cdot 225 \ , \ c_2(X_0) = 75 \cdot 235 \quad .$$

The A_4-singularities when resolved do not influence these values (cf. Brieskorn's theory on rational double points). Each of the remaining 50 singularities is resolved in a Fermat curve C of degree 5 with Euler number -10 and selfintersection number -5. The Milnor fibre of such a singularity has Euler number $1 + 4 \cdot 4 \cdot 4 = 65$. Therefore, each singularity reduces the Euler number by $65 + 10 = 75$. Each singularity reduces $c_1^2(X_0)$ by $-a^2 C^2$ where a is determined by the adjunction formula

$$a\,C\,C + C^2 = -e(C) = 10 \quad .$$

Thus $a = -3$ and $-a^2C^2 = 45$. Hence,

$$c_1^2(X) = 16875 - 50 \cdot 45 = 14625$$

$$c_2(X) = 17625 - 50 \cdot 75 = 13875 \quad .$$

There are 1875 A_4-configurations of (-2)-curves on X. We have (5)

$$m(A_4) = 15 - \frac{3}{5}$$

and

$$3c_2(X) - c_1^2(X) = 27000 = 1875 \cdot 14\frac{2}{5} \quad .$$

Hence, the equality sign holds in (6). It follows from Kobayashi's theorem in 4. that Y with the 50 Fermat curve singularities resolved equals Γ/B where Γ acts on the ball with isolated fixed points with isotropy groups of order 5.

The surface Y was originally constructed in a different way - in relation to the icosahedral line arrangement - by Th. Höfer in his Bonn dissertation [10].

6. LINE ARRANGEMENTS. We consider the complex projective plane $P_2(\mathbb{C})$ with homogeneous coordinates z_0, z_1, z_2. An arrangement of k lines is a set of k distinct lines in $P_2(\mathbb{C})$. They can be given by linear forms $\ell_1, \ldots, \ell_k$ in z_0, z_1, z_2. Let $t_r (r \geq 2)$ be the number of r-fold points, i.e., the number of points lying on exactly r lines of the arrangement. Then we have

$$\frac{k(k-1)}{2} = \sum_{r \geq 2} t_r \frac{r(r-1)}{2} \quad .$$

For an arrangement $\ell_1 = 0, \ldots, \ell_k = 0 (k \geq 3,\ t_k = 0)$ of lines we consider the function field

$$\mathbb{C}(z_1/z_0, z_2/z_0)((\ell_2/\ell_1)^{1/2}, \ldots, (\ell_k/\ell_1)^{1/2})$$

which is an abelian extension (Kummer extension) of the function field $\mathbb{C}(z_1/z_0, z_2/z_0)$ of $P_2(\mathbb{C})$ of degree 2^{k-1} and Galois group $(\mathbf{Z}/2\mathbf{Z})^{k-1}$. It determines an algebraic surface X with normal singularities which ramifies over the plane with the arrangement as locus of ramification. If the point $p \in P_2(\mathbb{C})$ lies on r lines of the arrangement $(r \geq 0)$, then there are 2^{k-1-r} points of X over p which are an orbit of the Galois group. For $r \geq 3$ these points are singular. The minimal resolution of such a point replaces the point by a smooth curve C of Euler number $2^{r-2}(4-r)$ and selfintersection number -2^{r-2}. We obtain a smooth surface Y associated to the arrangement.

Line arrangements and related algebraic surfaces were studied in [9]. Many more investigations were carried out in Th. Höfer's thesis [10] where he emphasized the relation to the work of P. Deligne and G.D. Mostow [3]. In this lecture we consider only some examples whose discussion we continue now.

As shown in [9] p. 132 we have

$$3c_2(Y) - c_1^2(Y) = 2^{k-3}(t_2+3t_3+t_4-k-\sum_{r\geq 5}(2r-9)t_r) \quad . \tag{7}$$

The surface Y contains $t_3 \cdot 2^{k-4}$ rational curves E_i of self-intersection number -2. They lie over the 3-fold points of the arrangement. The surface Y contains $t_4 \cdot 2^{k-5}$ elliptic curves C_j. They lie over the 4-fold points of the arrangement. We have $m(E_i) = \frac{9}{2}$ (see (5)) and $C_j^2 = -4$. Therefore,

$$\begin{aligned} &3c_2(Y) - c_1^2(Y) - \sum m(E_i) - \sum(-C_j^2) \\ &= 2^{k-3}(t_2 + \tfrac{3}{4}t_3 - k - \sum_{r\geq 5}(2r-9)t_r) \quad . \end{aligned} \tag{8}$$

It was shown in [9] that Y is of general type if $k \geqq 7$ and $t_k = t_{k-1} = t_{k-2} = t_{k-3} = 0$, it is of non-negative Kodaira dimension if $k \geqq 6$ and $t_k = t_{k-1} = t_{k-2} = 0$. In this talk, for simplicity, we always assumed general type. However, Miyaoka's inequality (6) is true if the surface has non-negative Kodaira dimension. From (6) and (8) we get

THEOREM. <u>For an arrangement of</u> k <u>lines in the complex projective plane we have</u>

$$t_2 + \frac{3}{4}t_3 \geq k + \sum_{r\geq 5}(2r-9)t_r \tag{9}$$

<u>provided</u> $t_k = t_{k-1} = t_{k-2} = 0$.

The inequality (9) is an improvement of an inequality mentioned in [9] p. 140. It does not seem to be known to experts in the theory of arrangements (see the literature quoted in [9]). There is an arrangement of 8 lines with $t_2 = 10$, $t_3 = 1$, $t_6 = 1$ for which (9) is wrong.

We mention some arrangements for which (9) is sharp and for

which the surface Y is related to the ball by the theorem in 4.

a) the nine inflection points of a smooth cubic surface determine the 12 lines of the Hesse arrangement with $k = 12$, $t_2 = 12$, $t_4 = 9$, $t_r = 0$ otherwise

b) the nine inflection points define 9 lines in the dual projective plane. This gives an arrangement with $k = 9$, $t_3 = 12$, $t_r = 0$ otherwise.

c) the simple group of order 168 operates on the complex projective plane (F. Klein [12]p. 101). It has 21 involutions. Each involution leaves a line pointwise fixed. We get an arrangement with $k = 21$, $t_3 = 24$, $t_4 = 21$ and $t_r = 0$ otherwise.

d) $(z_0^m - z_1^m)(z_1^m - z_2^m)(z_2^m - z_0^m) = 0$, $m \geq 4$, defines $3m$ lines with $t_2 = 0$, $t_3 = m^2$, $t_m = 3$ and $t_r = 0$ otherwise. We have the equality sign in (9) if and only if $m = 4$ or $m = 6$.

7. DOUBLE POINTS ON HYPERSURFACES. Let F_d be a smooth hypersurface of degree d in $P_3(\mathbb{C})$. It is easy to calculate

$$(10) \qquad 3c_2(F_d) - c_1^2(F_d) = 2d(d-1)^2 \quad .$$

Now we admit that F_d has ordinary nodes (double points) and is otherwise smooth. These nodes are points in whose neighborhood F_d can be given by $u^2 + v^2 + w^2 = 0$ with respect to a local analytic coordinate system of $P_3(\mathbb{C})$. If we pass to the minimal resolution $\tilde{F}_d$ of F_d , then each node is replaced by a smooth rational curve of selfintersection number -2 . By Brieskorn's theory $\tilde{F}_d$ belongs to the same "family" as F_d and (10) holds for $\tilde{F}_d$. The surface $\tilde{F}_d$ is of general type for $d \geq 5$ (it is a K3-surface, Kodaira dimension 0 , for $d = 4$). Let $\mu(d)$ be the maximum number of nodes on a hypersurface of degree d . For $d \geq 4$ we get by (5), (6) and (10)

$$(11) \qquad \mu(d) \leq \frac{4}{9} d\,(d-1)^2 \quad .$$

Miyqoka [17] discusses this inequality in relation to many classical and more recent results. (For example $\mu(5) = 31$ by A. Beauville.)

V.I. Arnold drew my attention to other results which are not mentioned by Miyaoka. See A. Varchenko [21]. S.V. Čmutov (cf.[21]) used the Čebyšev polynomials to define surfaces F_d with many nodes: The Čebyšev polynomial $T_d(x)$ is defined by

$$T_d(\cos\alpha) = \cos(d\alpha)$$

We have

$$T_d(x) = \sum (-1)^j \binom{d}{2j} x^{d-2j} (1-x^2)^j \quad .$$

The derivative $T_d'(x)$ has only simple zeros, they give the $d-1$ maxima and minima of $T_d(x)$. The maxima all have the value 1, the minima the value -1. We use affine coordinates x_1, x_2, x_3 for $P_3(\mathbb{C})$ and consider the Čmutov surface

$$T_d(x_1) + T_d(x_2) + T_d(x_3) = -1$$

which has no singularities on the infinite plane of $P_3(\mathbb{C})$. A node of the Čmutov surface has coordinates x_1, x_2, x_3 with $T_d'(x_i) = 0$, where among x_1, x_2, x_3 we must have two numbers for which T_d has a minimum. If $c(d)$ is the number of nodes, then

$$(12) \qquad \begin{aligned} c(d) &= 3 \cdot \left(\frac{d-1}{2}\right)^3 && \text{for } d \text{ odd} \\ c(d) &= 3 \cdot \left(\frac{d}{2}\right)^2 \left(\frac{d}{2} - 1\right) && \text{for } d \text{ even} \quad . \end{aligned}$$

By (11) and (12)

$$(13) \qquad \frac{3}{8} \leq \overline{\lim_{d\to\infty}}\, \mu(d)/d^3 \leq \frac{4}{9} \quad .$$

This seems to be all what is know about $\overline{\lim_{d\to\infty}}\, \mu(d)/d^3$.

8. SINGULARITIES ON PLANAR CURVES. Let C be a smooth curve of even degree $2k$ in $P_2(\mathbb{C})$. Let X_C be the 2-fold cover of the plane branched along C. Then

$$(14) \qquad 3c_2(X_C) - c_1^2(X_C) = k(10k-6) \ .$$

Now we admit that C has simple singularities. A singular point is simple if, with respect to a local analytic coordinate system, the curve can be given as follows

$A_k\ (k \geq 1)$	$x^2 + y^{k+1} = 0$
$D_k\ (k \geq 4)$	$y(x^2 + y^{k-2}) = 0$
E_6	$x^3 + y^4 = 0$
E_7	$x(x^2 + y^3) = 0$
E_8	$x^3 + y^5 = 0$.

Then X_C has singularities (rational double points) which blow up to configurations of (-2)-curves, namely to the configurations A_k, D_k, E_6, E_7, E_8 , on the smooth model $\tilde{X}_C$. For each simple singularity we define $m(p) = m(E)$ if p is blown up to the configuration E . We can apply (6) using that $\tilde{X}_C$ belongs to the same family as X_C by Brieskorn's theory and also satisfies (14).

THEOREM. Let C be a curve of degree $2k$ having only simple singularities $(k \geq 3)$. Then

$$\sum_{p \in \operatorname{sing}(C)} m(p) \leq k(10k-6) . \tag{15}$$

If d is the number of ordinary double points (A_1) and s the number of ordinary cusps (A_2) then by (15) and (5)

$$\frac{9}{2}d + 8s \leq k(10k-6) . \tag{16}$$

The paper of Ivinski's [11] contains many examples and references to the literature, in particular one finds a result of Varchenko [21]p. 164, related to (16). If $s(n)$ is the maximal number of cusps on a curve of degree n (with simple singularities only), then by (16)

$$\overline{\lim_{n\to\infty}} \frac{s(n)}{n^2} \leq \frac{5}{16} .$$

Varchenko (loc. cit.) obtains

$$\overline{\lim_{n\to\infty}} \frac{s(n)}{n^2} \leq \frac{23}{72} .$$

Examples show that this limes superior is not less than $\frac{1}{4}$.

The estimate

$$\frac{1}{4} \leqq \overline{\lim_{n\to\infty}} \frac{s(n)}{n^2} \leqq \frac{5}{16}$$

seems to be all what is known about it. Since a cusp reduces the class $n(n-1)$ of a smooth curve of degree n by 3, it is known classically that

$$\overline{\lim_{n\to\infty}} \frac{s(n)}{n^2} \leqq \frac{1}{3} \quad .$$

9. ARRANGEMENTS OF CONICS. Let $C_1,\ldots,C_k$ be distinct smooth conics in the complex projective plane $(k \geqq 4)$. We assume that the curve $C = C_1 \cup \ldots \cup C_k$ of degree $2k$ has only singularities of type A_1 or A_3 , which means that a point $p \in C$ lies on one or two conics, in the latter case p is a transversal intersection of the two conics or the two conics touch each other in p with intersection multiplicity 2 (tacnode). Let d be the number of transversal intersections and t the number of tacnodes. Then obviously,

$$d + 2t = 2k(k-1) \quad ,$$

whereas by (15) and (5)

$$\frac{9}{2}d + \frac{45}{4}t \leqq k(10k-6) \quad .$$

Eliminating d gives

$$t \leqq \frac{4}{9}(k^2 + 3k) \quad .$$

For $k = 4,5,6$ we get $t \leqq 12,17,24$. As U. Persson remarked there are 4 conics with $t = 12$ (and $d = 0$), there is a (projectively unique) arrangement of 5 conics with $t = 17$ given by I. Naruki [18]p. 1144, but the maximal t for $k \geqq 6$ is not known. There is a beautiful arrangement of 12 conics (Gerbaldi 1882; see Fricke-Klein [4]p. 648-649). The alternating group A_6 operates on the complex projective plane (Valentiner and Wiman) . There are six canonical subgroups of A_6 isomorphic to A_5 . They are icosahedral groups. Each leaves a conic fixed. Under an outer automorphism of

A_6 we get another system of six groups isomorphic to A_5 and again six conics. Each conic of the first system of six conics touches each conic of the second system in two points. We have $k = 12$ and $t = 72$, the above estimate gives $t \leq 80$.

A similar (classical) example with $k = 14$ and $t = 98$ can be obtained from the action of the simple group G_{168} on $P_2(\mathbb{C})$, see F. Klein [12] p. 106. There are two systems of seven subgroups of G_{168} isomorphic to the octahedral group S_4. Each subgroup leaves a conic invariant. Each conic of one system touches each concic of the other system in 2 points. Of the 98 tacnodes, 56 lie on the invariant curve f of degree 4, they are the 56 touching points of the 28 double tangents of f, and 42 lie on the invariant curve of degree 6.

For $k = 14$, the above estimate gives $t \leq 105$. I do not know whether $\overline{\lim\limits_{k\to\infty}} \frac{t(k)}{k^2}$ is positive ($t(k)$ = maximal number of tacnodes for an arrangement of k conics).

BIBLIOGRAPHY

[1] Bourguignon, J.-P.: Premières formes de Chern des variétés kähleriennes compactes [d'après E. Calabi, T. Aubin et S.T. Yau], Séminaire Bourbaki, Lecture Notes in Mathematics 710, Springer-Verlag Heidelberg (1979), 1-21.

[2] Brieskorn, E.: Rationale Singularitäten komplexer Flächen, Invent. math. 4 (1968), 336-358.

[3] Deligne, P. & Mostow, G.D.: Monodromy of hypergeometric functions and non-lattice integral monodromy, Preprint I.H.E.S. 1983.

[4] Fricke, R. & Klein, F.: Vorlesungen über die Theorie der automorphen Funktionen (Band 2), Nachdruck Teuber-Verlag Stuttgart 1965.

[5] Hemperly, J.C.: The parabolic contribution to the number of linearly independent automorphic forms on a certain bounded domain, Amer. J. Math. 94 (1972), 1078-1100.

[6] Hirzebruch, F.: Über vierdimensionale Riemannsche Flächen mehrdeutiger analytischer Funktionen von zwei komplexen Veränderlichen, Math. Ann. 126 (1953), 1-22.

[7] Hirzebruch, F.: Automorphe Formen und der Satz von Riemann-Roch, Symp. Int. Top. Alg. 1956, México Univ. México (1958), 129-144.

[8] Hirzebruch, F.: Hilbert's modular group of the field $\mathbb{Q}(\sqrt{5})$ and the cubic diagonal surface of Clebsch and Klein (in Russian), Uspekhi Mat. Nauk 31 : 5 (1976), 153-166; English translation in Russian Math. Surveys 31 : 5 (1976), 96-110.

[9] Hirzebruch, F.: Arrangements of lines and algebraic surfaces, Progress in Mathematics 36, Birkhäuser Boston (1983), 113-140.

[10] Höfer, Th.: Ballquotienten als verzweigte Überlagerungen der projektiven Ebene, Dissertation Bonn 1985.

[11] Ivinskis, K.: Normale Flächen und die Miyaoka-Kobayashi Ungleichung, Diplomarbeit Bonn 1985.

[12] Klein, F.: Über die Transformation siebenter Ordnung der elliptischen Funktionen, Ges. Math. Abhandlungen Band 3, LXXXIV, Reprint Springer-Verlag Berlin (1973), 90-136.

[13] Kobayashi, R.: Einstein-Kähler metrics on open algebraic surfaces of general type, Preprint (1984).

[14] Kobayashi, R.: Einstein-Kähler V-metrics on open Satake V-surfaces with isolated quotient singularities, Preprint (1984), to appear in Math. Ann.

[15] Miyaoka, Y.: On the Chern numbers of surfaces of general type, Invent. math. 42 (1977), 225-237.

[16] Miyaoka, Y.: Algebraic surfaces with positive indices, Progress in Mathematics 39, Birkhäuser Boston (1983), 281-301.

[17] Miyaoka, Y.: The maximal number of quotient singularities on surfaces with given numerical invariants, Math. Ann. 268 (1984), 159-171.

[18] Naruki, I.: Some invariants for conics and their applications, Publ. RIMS Koyoto Univ. 19 (1983), 1139-1151.

[19] Sakai, F.: Semi-stable curves on algebraic surfaces and logarithmic pluricanonical maps, Math. Ann. 254 (1980), 89-120.

[20] Van de Ven, A.: Some recent results on algebraic surfaces of general type, Séminaire Boubaki, Lecture Notes in Mathematics 677, Springer Verlag Heidelberg (1978), 155-166.

[21] Varchenko, A.N.: Asymptotics of integrals and Hodge structures (in Russian), Itogi Nauki i Techniki, Series "Contemporary Problems of Mathematics" Vol. 22 Moscow (1983), 130-166.

[22] Yau, S.T.: Calabi's conjecture and some new results in algebraic geometry, Proc. Nat. Acad. Sci. USA 74 (1977), 1798-1799.

Max-Planck-Institut für Mathematik
Gottfried-Claren-Straße 26
D-5300 Bonn 3

Federal Republic of Germany

Contemporary Mathematics
Volume 58, Part I, 1986

METRICS ON MODULI SPACES

N.J. Hitchin

ABSTRACT. A hyperkähler metric is shown to exist on an open set of the moduli space of all flat bundles on a Riemann surface. The moduli space of irreducible unitary connections together with its natural metric is the fixed point set of a circle action in this space. A close analogy with the moduli space of vortices is described, in the context of hyperkählerian moment maps.

1. INTRODUCTION. The notion of moduli space runs through a large part of mathematics, from algebraic geometry to mathematical physics and even, with the work of Donaldson, in differential topology. A moduli space is the set of equivalence classes of certain geometrical objects modulo the action of a large group of natural transformations, for example the original moduli space: the space of all complex structures on a Riemann surface modulo the group of diffeomorphisms. These spaces have been studied in the past from the point of view of their topology or complex structure but in many cases they have, as a consequence of their geometrical origin, a natural Riemannian metric which has been largely ignored. The purpose of this lecture is to consider two such moduli spaces, one from algebraic geometry and one from mathematical physics, from a metrical point of view. Both

are well understood from a topological point of view, and both are equipped with natural metrics. We shall see that these metrics are derived from a very special branch of differential geometry - hyperkähler geometry - and that this fact raises the possibility of finding the metrics explicitly.

2. BUNDLES ON RIEMANN SURFACES. Let X be a compact Riemann surface of genus $g > 1$. The first moduli space we shall consider is the space of equivalence classes of stable holomorphic vector bundles of rank n and degree k over X. This generalizes the Jacobian of X, which is the moduli space of line bundles of degree zero. The Jacobian has a flat metric obtained from the natural hermitian form on $H^1(X,\underline{O})$: using Hodge theory we can represent any element by a unique closed (and hence harmonic) $(0,1)$ form α, and then define

$$(\alpha,\beta) = \frac{1}{i}\int_X \alpha \wedge \overline{\beta} .$$

For bundles of higher rank, we need to use the theorem of Narasimhan and Seshadri [9]: a bundle is stable if and only if it arises from an irreducible projective unitary representation of the fundamental group of X. In other words, a holomorphic bundle is stable if and only if it admits a unitary connection A whose curvature is central. In the case of degree 0, this is simply a flat unitary connection. In either case, the covariant derivative defines an elliptic complex:

$$\Omega^0(\mathrm{End}\ E) \xrightarrow{d_A} \Omega^1(\mathrm{End}\ E) \xrightarrow{d_A} \Omega^2(\mathrm{End}\ E) .$$

Now the tangent space to the moduli space is identified with the sheaf cohomology group $H^1(X,\underline{0}(\mathrm{End}\ E))$ and as before, we can represent this by a unique d_A-closed (0,1) form and define

$$(\alpha,\beta) = \frac{1}{i}\int_X \mathrm{Trace}(\alpha \wedge \beta^*) .$$

The connection A is uniquely determined modulo unitary automorphisms of the bundle (gauge transformations) which leave the metric unchanged, so this gives a well-defined metric on the moduli space.

The metric (introduced by Narasimhan) gives an obvious generalization of the flat metric which defines the polarization of the Jacobian of a curve, and hence would be expected to carry information about the structure of the curve, but it is unknown in even the simplest case. This contrasts with the extensive information available concerning the topological and analytical structure [1]. For example, the moduli space of stable rank 2 bundles of fixed determinant and degree 1 on a curve of genus 2 is the complete intersection of two quadrics in $\mathbb{P}^5$.

Let us turn now to an example from mathematical physics.

3. VORTICES. Let L be a complex line bundle on $\mathbb{R}^2$ with a unitary connection A, and a section Φ of L^2. We consider the functional

$$\mathcal{A} = \int_{\mathbb{R}^2} |F_A|^2 + |D_A\Phi|^2 + \tfrac{1}{2}(1 - |\Phi|^2)^2$$

where F_A is the curvature of the connection and D_A is the covariant derivative of Φ. This two-dimensional gauge theory is called the Abelian Higgs model, and was introduced by Ginzburg and Landau in 1950 in the study of superconductivity. In order to make the action $\mathcal{A}$ converge we need (A,Φ) to have the following behaviour:

$$|\Phi| \to 1$$

$$|D_A\Phi| \to 0 \qquad |F_A| \to 0, \qquad \text{as} \quad |x| \to \infty .$$

The first condition implies the existence of a map from a large circle in $\mathbb{R}^2$ to the unit circle which has a degree, k. If the limits above are sufficiently uniform, then an integration by parts gives a lower bound to the action:

$$\mathcal{A} \geq 2\pi|k|$$

and furthermore this absolute minimum is achieved when (A,Φ) satisfy the coupled equations:

$$\left.\begin{aligned} \bar{\partial}_A\Phi &= 0 \\ F_A &= \tfrac{1}{2}(|\Phi|^2 - 1) \end{aligned}\right\} \tag{3.1}$$

Here $\bar{\partial}$ denotes the $(0,1)$ part of the covariant derivative. The first equation in (3.1) therefore says that Φ is holomorphic with respect to the holomorphic structure on L defined by the connection A. Solutions to these equations are called vortices [6].

Any unitary automorphism of L (gauge transformation) clearly takes one solution to another, and the moduli space consists of the space of all solutions modulo gauge equivalence. The basic result on the structure of this is the existence theorem of Jaffe and Taubes [6], that given k points x_i in $\mathbb{R}^2$ (or points with total multiplicity k), there exists a solution to the vortex equations, unique modulo gauge equivalence, with $\Phi(x_i) = 0$ and $\mathcal{A} = 2\pi|k|$. This means that the moduli space is $\mathbb{R}^{2k} = \mathbb{C}^k$: the coefficients of the complex polynomial whose roots are $x_i \in \mathbb{R}^2 = \mathbb{C}$. Thus the topological structure of the moduli space is very simple. An explicit form for the solutions to (3.1) is, however, unknown.

To produce a metric on the moduli space we linearize the equations (3.1) to produce linear equations for $(\dot{A},\dot{\Phi})$ and then impose the condition $d_A^*\dot{A} = 2\mathrm{Im}\bar{\Phi}\dot{\Phi}$. Setting up a suitable elliptic boundary value problem for this set of equations will lead to uniqueness of this representative and the metric is then defined as

$$\int_{\mathbb{R}^2} (\dot{A},\dot{B}) + (\dot{\Phi},\dot{\Psi}) \ .$$

There are analytical difficulites in carrying out this programme for vortices, so we shall concentrate on the first example of stable bundles on Riemann surfaces. However, it will become evident that the special properties of the metrics are consequences of the equations which define the moduli spaces, and these are common to both examples given here. It should be mentioned that one of the goals in studying the metric for vortices is in the "scattering" behaviour, as described for magnetic monopoles in [3].

4. MOMENT MAPS. The setting for the special metrics we shall investigate is that of a symplectic manifold with a group action. We recall the geometry of this situation next.

Let M^{2n} be a manifold with a symplectic form ω and suppose the Lie group G acts on M, preserving ω. If X is a vector field generated by this action, then the Lie derivative $\mathcal{L}_X\omega$ vanishes. Now for any differential form

$$\mathcal{L}_X\omega = i(X)d\omega + d(i(X)\omega)$$

hence

$$0 = d(i(X)\omega)$$

and so (if $H^1(M,\mathbb{R}) = 0$) there exists a function $\mu_X : M \to \mathbb{R}$ such that

$$d\mu_X = i(X)\omega \ .$$

As X ranges over the Lie algebra of G, we can put all these functions together to obtain a map

$$\mu : M \to \mathfrak{g}^* \qquad (4.1)$$

to the dual of the Lie algebra defined by

$$\langle\mu(x),X\rangle = \mu_X(x) \ .$$

There is a natural action of G on both sides of (4.1) and a constant ambiguity in the choice of μ_X. If these can be adjusted to make μ commute with the action of G, then μ is called a moment map. The remaining ambiguity in the definition of μ is the addition of a constant abelian character in $\mathfrak{g}^*$. Suppose G acts freely and discontinuously, then

$$\mu^{-1}(0)/G$$

is a <u>symplectic manifold of dimension $2n - 2 \dim G$</u>. This is the Marsden-Weinstein quotient of a symplectic manifold by a group. More generally, if $\alpha \in \mathfrak{g}^*$ and G_α denotes the stabilizer of α, then

$$\mu^{-1}(\alpha)/G_\alpha$$

is also a symplectic manifold (see [7], for example).

Suppose now that M is a Kähler manifold i.e. a manifold endowed with a metric g and a compatible complex structure I which is covariant constant with respect to the Levi-Civita connection. Then

$$\omega(X,Y) = g(IX,Y)$$

defines a symplectic form, the <u>Kähler form</u>, on M. If the group G now acts as isometries as well as preserving the symplectic form, then the quotient $\mu^{-1}(0)/G$ has both a symplectic structure and a metric. These are related by the following

(4.2) PROPOSITION: <u>The induced metric on $\mu^{-1}(0)/G$ is Kählerian.</u>

PROOF: The metric we want is produced by first taking the induced metric on the submanifold $\mu^{-1}(0) \subset M$ and then taking the quotient metric on $\mu^{-1}(0)/G$. The quotient metric is the inner product induced on the vectors perpendicular to the orbits of G in $\mu^{-1}(0)$, which is identified via the projection with the tangent bundle of $\mu^{-1}(0)/G$.

Now the Levi-Civita connection of the submanifold $\mu^{-1}(0)$ is obtained by orthogonal projection from the tangent space of M to the tangent space of $\mu^{-1}(0)$. Furthermore, the Levi-Civita connection of $\mu^{-1}(0)/G$ induces a connection on the horizontal tangents in $\mu^{-1}(0)$. This is in fact again obtained by projection. This is seen to be so if we take two commuting vector fields X, Y on $\mu^{-1}(0)/G$ and lift them horizontally to $\tilde{X}$, $\tilde{Y}$ on $\mu^{-1}(0)$. They no longer commute, because $\mu^{-1}(0)$ is a principal G-bundle over $\mu^{-1}(0)$ with a connection defined by the horizontal tangent vectors. However,

$$[\tilde{X},\tilde{Y}] = F(X,Y)$$

where F is the curvature of this connection. Since this is vertical,

$$\nabla_{\tilde{X}} \tilde{Y} - \nabla_{\tilde{Y}}\tilde{X} = [\tilde{X},\tilde{Y}] = F(X,Y)$$

and this projects to zero in the horizontal direction, so the projected connection is torsion-free. It clearly preserves the metric and so coincides with the Levi-Civita connection.

Consider next the orthogonal complement of the horizontal space in TM restricted to $\mu^{-1}(0)$. This is spanned by vector fields $X_1,\ldots,X_m$ arising from a basis of $\mathfrak{g}$ and the normal vectors to $\mu^{-1}(0)$, $\operatorname{grad}\mu_{X_1},\ldots,\operatorname{grad}\mu_{X_m}$. But

$$g(\operatorname{grad}\mu_X,Y) = d\mu_X(Y) = \omega(X,Y) = g(IX,Y)$$

so

$$\operatorname{grad}\mu_X = IX. \tag{4.3}$$

Thus this space, and hence the horizontal space itself, is complex.

Now, since the metric on M was Kählerian, I commutes with the covariant derivative, and since the projection onto the horizontal space commutes with I, so does the Levi-Civita connection on $\mu^{-1}(0)/G$ which is therefore Kählerian as required.

The Kähler form is in fact the symplectic form of the Marsden-Weinstein quotient.

This infinitesimal result disguises a more geometrical interpretation: since the tangent bundle of $\mu^{-1}(0)/G$ is the quotient of the tangent bundle of M by the <u>complexification</u> of the vectors generated by $\mathfrak{g}$, we might expect that, as a complex manifold,

$$\mu^{-1}(0)/G \cong M/G^c \tag{4.4}$$

where G^c denotes the complexificaiton of the Lie group G. In many cases, this turns out to be so, provided one restricts attention to the open set of stable points in M (see [7]).

This point of view led Atiyah & Bott [1] to consider an infinite dimensional version, which leads to the natural metric on the moduli space of stable bundles on a Riemann surface X.

Let $\mathcal{A}$ be the space of $\bar{\partial}$-operators on a C^∞ hermitian vector bundle E over X. This may be thought of as either the space of all complex structures on E, which is what a $\bar{\partial}$-operator is, or the space of all unitary connections. It is an infinite dimensional complex affine space with a Kähler metric:

$$(\alpha,\beta) = \frac{1}{i}\int_X \mathrm{Tr}(\alpha \wedge \beta^*) \qquad (\alpha,\beta \in \Omega^{0,1}(\mathrm{End}\ E)).$$

The group $\mathcal{G}$ of unitary gauge transformations acts on $\mathcal{A}$ preserving the metric and Kähler form. The moment map for this action is

$$\mu : \mathcal{A} \to \Omega^2(\text{End } E)$$

$$\mu(A) = F_A = \text{curvature of the connection } A.$$

Thus $\mu^{-1}(0)/\mathcal{G}$ is the space of equivalence classes of flat unitary connections. The space $\mathcal{A}/\mathcal{G}^c$ is the space of equivalence classes of complex structures on E. The analogue of (4.4) is the theorem of Narasimhan and Seshadri, and the natural Kähler metric on $\mu^{-1}(0)/G$ coming from the quotient construction described above is the one constructed in §2. This is because $d_A\alpha = 0$ if and only if α is orthogonal to the orbits of $\mathcal{G}$.

A proof of the theorem of Narasimhan and Seshadri along these lines was given by Donaldson [4].

We see then that the natural metric on the moduli space of stable bundles is obtained by a quotient construction from a flat Kähler manifold. We shall see later that it sits inside a more special type of metric space produced by an analogous construction - a <u>hyperkähler manifold</u>.

5. HYPERKÄHLER METRICS. A <u>hyperkähler manifold</u> is a manifold M^{4n} with a Riemannian metric g compatible with three complex structures I,J,K which are covariant constant with respect to the Levi-Civita connection and satisfy the quaternionic identities:

$$I^2 = J^2 = K^2 = -1 \qquad IJ = -JI = K \quad \text{etc.}$$

Equivalently, a hyperkähler metric is a metric with holonomy contained in $Sp(n)$.

Corresponding to the three complex structures, there are three Kähler forms ω_1, ω_2 and ω_3. These three symplectic forms are themselves sufficient to determine the metric, so we might equally call M hypersymplectic.

Suppose G is a Lie group of isometries acting on M and preserving the three symplectic forms, then we obtain <u>three</u> moment maps μ_1, μ_2 and μ_3, or a single vector-valued moment map

$$\mu : M \to \mathfrak{g}^* \otimes \mathbb{R}^3.$$

The main result we shall use (due to the author, Roček, Karlhede & Lindström [5]) is:

(5.1) THEOREM: <u>Let G be a Lie group acting freely on a hyperkähler manifold M and preserving the hyperkähler structure. Then the induced metric on $\mu^{-1}(0)/G$ is hyperkählerian</u>.

PROOF: Fix first one complex structure I, and consider the complex function

$$\nu = \mu_2 + i\mu_3 : M \to \mathfrak{g}^* \otimes \mathbb{C}.$$

For each vector field X generated by $\mathfrak{g}$, we have

$$\begin{aligned} d\nu_X(Y) &= \omega_2(X,Y) + i\omega_3(X,Y) \\ &= g(JX,Y) + ig(KX,Y) \end{aligned}$$

and

$$d\nu_X(IY) = g(JX,IY) + ig(KX,IY)$$
$$= -g(KX,Y) + ig(JX,Y) \ .$$

Hence

$$d\nu_X(IY) = id\nu_X(Y) \qquad \text{i.e.} \quad \bar{\partial}\nu_X = 0 \ .$$

Thus ν is holomorphic, with respect to I. It follows that $\nu^{-1}(0)$ is a complex submanifold of a Kähler manifold and hence its induced metric is Kählerian. The group G acts on $\nu^{-1}(0)$ preserving the Kähler form and its moment map is clearly μ_1. Hence by (4.2) the quotient metric on

$$\nu^{-1}(0) \cap \mu_1^{-1}(0)/G = \mu^{-1}(0)/G$$

is Kählerian with respect to the complex structure I.

To complete the proof, repeat with the complex structures J and K.

Note that the function ν is the moment map for the complexification of G with respect to the holomorphic symplectic form $\omega_2 + i\omega_3$. Holomorphic, that is, with respect to I. Thus, in the spirit of (4.4), each complex structure on $\mu^{-1}(0)/G$ is the holomorphic Marsden-Weinstein quotient of M by G^c. The manifold $\mu^{-1}(0)/G$ clearly has dimension $4n - 4 \dim G$.

The aspect of hyperkähler geometry which makes this construction interesting is the special nature of such metrics. In particular, the Ricci tensor of a hyperkähler metric vanishes and so we obtain solutions to Einstein's vacuum equations. Indeed in four dimensions, a hyperkähler metric is the same thing as a self-dual Einstein metric with zero scalar curvature. Theorem (5.1) in fact generates non-trivial solutions from trivial ones, as the following example shows.

EXAMPLE: Let V be a hermitian vector space and set

$$M = V \times V^* .$$

This is a flat hyperkähler manifold.
Let G be $U(1)$ acting via

$$e^{i\theta}(v,\alpha) = (e^{i\theta}v, e^{-i\theta}\alpha),$$

then G preserves the metric, the hermitian form and the complex symplectic form $\langle v_1,\alpha_2\rangle - \langle v_2,\alpha_1\rangle$ and hence the hyperkähler structure. Since G is abelian we have a choice of moment map and we take

$$\begin{aligned} \mu_1(v,\alpha) &= \|v\|^2 - \|\alpha\|^2 - 1 \\ \mu_2 + i\mu_3(v,\alpha) &= \langle v,\alpha\rangle \end{aligned} \tag{5.2}$$

Now $G^c \cong \mathbb{C}^*$ acts on $\{(v,\alpha) \mid \langle v,\alpha\rangle = 0\}$ by $\lambda\cdot(v,\alpha) = (\lambda v, \lambda^{-1}\alpha)$ and we can choose λ to make

$$\|\lambda v\|^2 - \|\lambda^{-1}\alpha\|^2 - 1 = 0$$

by setting $|\lambda|^2 = \dfrac{1+\sqrt{1+4\|v\|^2\|\alpha\|^2}}{2\|v\|^2}$ and this is unique modulo $U(1)$. Clearly if $v = 0$, there is no solution for any λ.

Thus in each $\mathbb{C}^*$-orbit with $v \neq 0$ there is a point in $\mu_1^{-1}(0)$, unique modulo $U(1)$. Hence:

$$\begin{aligned} \mu^{-1}(0)/G &\cong \{(\alpha,v) : \langle\alpha,v\rangle = 0 \ \& \ v \neq 0\}/\mathbb{C}^* \\ &= T^*P(V). \end{aligned}$$

That is, the standard complex structure on $V \times V^*$ induces the complex structure of the cotangent bundle of complex projective space on $\mu^{-1}(0)/G$. The hyperkähler metric one obtains on the quotient is the one discovered by Calabi. When $\dim V = 2$, it is the Eguchi-Hanson metric.

5. MODULI OF FLAT CONNECTIONS. We consider now the infinite-dimensional version of this example, modelled on the Atiyah-Bott approach to stable bundles in §4. Again, let X be a compact Riemann surface, E a C^∞ hermitian vector bundle and $\mathcal{A}$ the space of $\bar\partial$-operators. We set

$$M = \mathcal{A} \times \bar{\mathcal{A}}$$

where $\bar{\mathcal{A}}$ is the space of all ∂-operators. The infinite-dimensional affine space M may be considered as the <u>space of all (not necessarily unitary) connections on E, by setting $\partial + \bar\partial = d_A$, the covariant derivative</u>. The group of unitary gauge transformations $\mathcal{G}$ acts on M preserving the hyperkähler structure, and so we have three moment maps:

$$\begin{aligned} \mu_1(\bar\partial,\partial) &= F_1 - F_2 \\ \mu_2 + i\mu_3(\bar\partial,\partial) &= F = \text{curvature of } d_A . \end{aligned} \tag{5.3}$$

Here F_1 is the curvature of the unique unitary connection compatible with $\bar\partial$ and F_2 is the curvature of the unique unitary connection compatible with ∂.

The zero set of the complex moment map $\mu_2 + i\mu_3$ is the space of all flat $GL(n,\mathbb{C})$ connections, and following (4.4) we expect to obtain a hyperkähler metric on the space of flat connections modulo complex gauge equivalence, which is $\mathrm{Hom}(\pi_1(X), GL(n,\mathbb{C}))/GL(n,\mathbb{C})$. We must of course restrict to those connections in whose orbit there is a zero of the moment map μ_1, which is the analogue of the condition $v \neq 0$ in the above example of $T^*P(V)$.

We shall treat elsewhere the question of which flat connections are stable in the above sense. It is enough for our purposes to find an open set for which the equations (5.3) have a solution.

First note that if $\partial + \bar{\partial}$ is a flat <u>unitary</u> connection, then $F_1 = F_2 = 0$ and we have a solution.

Next consider the linearization of (5.3) at a flat unitary connection. If we set

$$\dot{\bar{\partial}} = \dot{A} \in \Omega^{0,1}(\mathrm{End}\ E)$$

$$\dot{\partial} = \dot{B} \in \Omega^{1,0}(\mathrm{End}\ E)$$

then we obtain

$$d\mu_1(\dot{A},\dot{B}) = *\mathrm{Im}((\partial + \bar{\partial})*(\dot{A},\dot{B})) \in \Omega^2(\mathrm{End}\ E)$$

$$d\mu_2(\dot{A},\dot{B}) = (\partial + \bar{\partial})(\dot{A},\dot{B}) \in \Omega^2(\mathrm{End}\ E) \otimes \mathbb{C}\ .$$

A unitary gauge transformation ψ gives

$$(\dot{A},\dot{B}) = (\bar{\partial}\psi, \partial\psi)$$

hence

$$\frac{\bigcap_{i=1}^{3} \ker d\mu_i \subset \Omega^1(\mathrm{End}\ E)\otimes\mathbb{C}}{\mathrm{im}\,\partial+\bar{\partial}:\Omega^0(\mathrm{End}\ E)\to\Omega^1(\mathrm{End}\ E)\otimes\mathbb{C}}$$

is isomorphic to the harmonic 1-forms in the elliptic complex:

$$\Omega^0(\mathrm{End}\ E) \otimes \mathbb{C} \to \Omega^1(\mathrm{End}\ E) \otimes \mathbb{C} \to \Omega^2(\mathrm{End}\ E) \otimes \mathbb{C}$$

for the flat connection $\partial + \bar{\partial} = d_A$.

If the flat unitary connection at which we are linearizing is irreducible, then standard deformation arguments [2, 8] show that (5.3) can be solved for nearby flat connections. We then obtain a hyperkähler metric on the $n^2(4g-4)$ dimensional space of equivalence classes of flat connections, in a neighbourhood of the flat unitary ones.

Note that we have described only one of the complex structures of the hyperkähler metric, arising from the product complex structure on $M = \mathcal{A} \times \bar{\mathcal{A}}$. If we denote by $\partial_1 + \bar{\partial}_1$ the unitary connection corresponding to $\bar{\partial}_1 \in \mathcal{A}$ and $\partial_2 + \bar{\partial}_2$ the unitary connection compatible with $\partial_2 \in \bar{\mathcal{A}}$, then the isomorphism

$$(\bar{\partial}_1, \partial_2) \to (\tfrac{1}{2}(\bar{\partial}_1 + \bar{\partial}_2), \tfrac{1}{2}(\partial_2 - \partial_1))$$

from $\mathcal{A} \times \bar{\mathcal{A}}$ to $\mathcal{A} \times \Omega^{1,0}(\mathfrak{g})$ defines another constant complex structure on $M = \mathcal{A} \times \bar{\mathcal{A}}$. This time it is a product of a complex affine space with a complex vector space, so the group of complex gauge transformations act differently.

Let us put $\bar{\partial} = \tfrac{1}{2}(\bar{\partial}_1 + \bar{\partial}_2)$, $\phi = \tfrac{1}{2}(\partial_2 - \partial_1)$ then with respect to this complex structure, the moment maps are:

$$\left.\begin{aligned} \mu_1(\bar{\partial},\phi) &= F + \tfrac{1}{2}[\phi,\phi^*] \\ \mu_2 + i\mu_3(\bar{\partial},\phi) &= \bar{\partial}\phi \end{aligned}\right\} . \qquad (5.4)$$

In this form the equations clearly admit a $U(1)$ action:

$$\phi \to e^{i\theta}\phi .$$

It is an action which only fixes this complex structure (and its conjugate) and so is not an obvious action on the space of flat connections - it is determined by the metric there rather than the complex structure. Let us consider the fixed points of this action on the moduli space of solutions to (5.4).

A fixed point will be given by a pair $(\bar\partial,\phi) \in \mathcal{A} \times \Omega^{1,0}(\mathfrak{g})$ for which there exist unitary gauge transformations $u(\theta)$ such that

$$\left.\begin{aligned} u(\theta)^{-1}\phi u(\theta) &= e^{i\theta}\phi \\ \bar\partial u(\theta) &= 0 \end{aligned}\right\} \qquad (5.5)$$

The second of these equations implies that, if $u(\theta)$ is not a constant scalar, the connection is reducible. In a neighbourhood of the irreducible flat connection on which our hyperkähler metric is defined, this is impossible, hence the only fixed points there are where $\phi = e^{i\theta}\phi$ i.e. $\phi = 0$. Then $F = 0$ and we have the moduli space of flat irreducible unitary connections.

Consequently the moduli space of flat unitary connections (the moduli space of stable bundles) is, with its canonical metric, the fixed point set of a group of isometries and is thus a totally geodesic submanifold of a hyperkähler manifold.

6. MODULI OF VORTICES. We shall consider next the same set-up for $SU(2)$ connections on $\mathbb{R}^2$. We set

$$M = \mathcal{A} \times \Omega^{1,0}(\mathfrak{g})$$

where $(\bar\partial,\phi)$ differ from some standard pair by suitably decaying forms.

The equations (5.4) in this setting appear to have no non-trivial solutions. (We may point out that they are in fact the self-duality equations in $\mathbb{R}^4$ which are invariant under an $\mathbb{R}^2$ translation group). Thus in the moment map formalism $\mu^{-1}(0)$ may be empty.

Consider however $F_0 = \begin{pmatrix} \frac{1}{2}i & 0 \\ 0 & -\frac{1}{2}i \end{pmatrix} \in \Omega^2(\mathfrak{g})$.

The subgroup of $SU(2)$ gauge transformations which preserve F_0 consists of $U(1)$ gauge transformations, $\mathcal{G}_0$. From the quotient construction we expect then to obtain a hyperkähler metric on

$$\mu^{-1}(F_0, 0, 0)/\mathcal{G}_0 .$$

This is the moduli space of solutions to the equations

$$\left.\begin{aligned} &\bar{\partial}\phi = 0 \\ &F + \tfrac{1}{2}[\phi, \phi^*] = F_0 \end{aligned}\right\} \qquad (6.1)$$

If we again look at the circle action $\phi \to e^{i\theta}\phi$, then at a fixed point in the moduli space the $SU(2)$ connection defined by $\bar{\partial}$ will reduce to a $U(1)$ connection A with curvature

$$F = \begin{pmatrix} F_A & 0 \\ 0 & -F_A \end{pmatrix}$$

and, relative to this decomposition, we must have

$$\phi = \begin{pmatrix} 0 & \Phi dz \\ 0 & 0 \end{pmatrix}$$

according to the first equation of (5.5).

The equations (6.1) then become

$$\left.\begin{aligned} \bar{\partial}_A \Phi &= 0 \\ F_A &= \tfrac{1}{2}(|\Phi|^2 - 1) \end{aligned}\right\}$$

which are the vortex equations (3.1). Here, from the work of Jaffe and Taubes, we have non-trivial solutions but as yet the appropriate elliptic estimates which will give a good deformation theory to produce solutions of the general equations (6.1) for non-abelian vortices have not been worked out.

Nevertheless, formally speaking, we expect that the moduli space of vortices should be the fixed point set of an isometric circle action and hence a <u>totally geodesic submanifold of a hyperkähler manifold</u>.

7. TWISTOR THEORY. Why, one might ask, are hyperkähler metrics so interesting? In the examples we have given the hyperkähler metric exists on a manifold twice the dimension of the original manifold we were considering and there may well be a theorem that any analytic Kähler manifold can be embedded as the fixed point set of a circle in a hyperkähler manifold. This would not say anything special about the Kähler metric itself. The answer is not so much that the <u>local</u> metric structure of moduli spaces is special, but that the hyperkähler metrics in which they sit arise in a natural way and moreover, thanks to twistor theory, have a <u>holomorphic</u> description. This raises the hope that it may be possible to go directly from holomorphic data - stable bundles on a Riemann surface - to the hyperkähler metric.

We shall here describe briefly the twistor approach. It is a direct generalization of Penrose's non-linear graviton construction [10] in a context developed by Salamon [11]. Details may be found in [5].

If M^{4n} is a hyperkähler manifold, then it has by definition three covariant constant complex structures I, J, K. In fact any linear combination $aI + bJ + cK$ where $a^2 + b^2 + c^2 = 1$ is also a complex structure, so there is a family of complex structures parametrized by S^2. Using the standard complex structure on S^2, these define a complex structure on the <u>twistor space</u> $Z = M \times S^2$ which is moreover integrable, and fibres holomorphically over $S^2 = \mathbb{P}^1$. The three Kähler forms combine to give a holomorphic 2-form $(\omega_2 + i\omega_3) + 2i\zeta\omega_1 + (\omega_2 - i\omega_3)\zeta^2$ along the fibres of $Z \to \mathbb{P}^1$, but with values in the line bundle $\underline{O}(2)$. The twistor space has a real structure defined by the antipodal map on S^2. The copy of $\mathbb{P}^1$ defined by a point $x \in M$ in the product $M \times S^2 = Z$ is holomorphic and real.

The force of the twistor construction is the converse theorem:

<u>THEOREM</u> [5]: Let Z^{2n+1} be a complex manifold with

(i) a fibration $p : Z \to \mathbb{P}^1$,

(ii) a family of sections with normal bundle $\mathbb{C}^{2n}(1)$,

(iii) a holomorphic section $\omega \in \Omega^2_{Z/\mathbb{P}^1}(2)$ defining a symplectic form on each fibre,

(iv) a real structure τ compatible with (i), (ii) and (iii) and inducing the antipodal map $\zeta \to -\bar{\zeta}^{-1}$ on $\mathbb{P}^1$.

Then, the family of real sections possesses a natural hyperkähler metric for which Z is the twistor space.

From an algebraic geometric point of view Z is a symplectic manifold defined over the field of rational functions in one variable and the sections of $Z \to \mathbb{P}^1$ are rational points. The quotient construction of §5 is then, on the twistor space level, the Marsden-Weinstein quotient within this category.

Twistor methods have the advantage of making constructions look natural, but in actually finding the twistor space a number of problems appear. One of these is that the twistor space for flat connections on a Riemann surface is non-Kähler. This is a feature which is shared by the Calabi metric described in §5. Depite that, it may still be algebraic and some sequence of algebraic processes from the starting point of holomorphic bundles to the end point of the twistor space Z may possibly succeed in producing the natural metric on the moduli space of stable bundles.

BIBLIOGRAPHY

[1] M.F. Atiyah & R. Bott, "The Yang-Mills equations over Riemann surfaces", Phil. Trans. Roy. Soc. London A308 (1982), 523-615.

[2] M.F. Atiyah, N.J. Hitchin & I.M. Singer, "Self-duality in four-dimensional Riemannian geometry", Proc. Roy. Soc. London A362 (1978), 425-461.

[3] M.F. Atiyah and N.J. Hitchin, "Low energy scattering of non-abelian monopoles", Phys. Lett. 107A (1985), 21-25.

[4] S.K. Donaldson, "A new proof of a theorem of Narasimhan and Seshadri", J. Differential Geometry 18 (1983), 269-277.

[5] N.J. Hitchin, A. Karlhede, U. Lindström & M. Roček, "Algebraic constructions of hyperkähler manifolds", (to appear).

[6] A. Jaffe & C. Taubes,"Vortices and monopoles", Birkhäuser, Boston (1980).

[7] F.C. Kirwan, "Cohomology of quotients in algebraic and symplectic geometry",Princeton Mathematical Notes 31, Princeton Univ. Press, Princeton (1985).

[8] M. Kuranishi, "New proof for the existence of locally complete families of complex structures", in Proc. of the Conference on Complex Analysis (ed. A. Aeppli et al) Minneapolis (1964), Springer Verlag, New York (1965).

[9] M.S. Narasimhan & C.S. Seshadri, "Stable and unitary vector bundles on a compact Riemann surface", Ann. of Math., 82 (1965), 540-564.

[10] R. Penrose, "Nonlinear gravitons and curved twistor theory", Gen. Relativity Gravitation 7, (1976), 31-52.

[11] S. Salamon, "Quaternionic Kähler manifolds", Invent. Math. 67, (1982), 143-171.

Mathematical Institute,
24-29 St. Giles,
Oxford OX1 3LB,
England.

Contemporary Mathematics
Volume 58, Part I, 1986

VARIETIES WITH RATIONAL SINGULARITIES

George R. Kempf*

The theory of rational singularities of varieties of dimension greater than two has developed quite nicely since its beginnings in [2,4]. In this note we shall survey some of this territory.

Let $f : X \to Y$ be a proper birational morphism between varieties where X is smooth. By duality theory we have a trace homomorphism $f_*\omega_X \to \omega_Y$ where ω_* denotes the dualizing sheaf of $*$. The sheaf $f_*\omega_X$ is independent of the resolution and it will be called the sheaf of absolutely regular differentials $\equiv K_Y$. Thus K_Y consists of those dualizing differentials which acquire no poles when pulled by birational morphisms.

In characteristic zero we can define rational singularities as follows. A variety Y has rational singularities if

a) Y is Cohen-Macaulay and

b) $K_Y = \omega_Y$; i.e., every dualizing differential is absolutely regular.

A smooth variety has rational singularities.

We will study when a hypersurface has rational singularities. Let H be an irreducible divisor on a smooth variety A. Using Hironaka's resolution theorem we may achieve the following situation. We have a proper birational morphism $g : A' \to A$ where A' is another smooth variety such that $g^{-1}(H) = H' + \Sigma m_i E_i$ where H' and the E_i's are smooth divisors meeting with transversal intersections and $f = g|_{H'} : H' \to H$ is birational.

Thus the integer m_i tells how many times a local equation for H pulled-back to A' vanishes on the exceptional divisors E_i. Similarly we can let n_i be the number of times a local basic differential on A vanishes along E_i. Then we have a canonical isomorphism

$$g^*\omega_A(\Sigma n_i E_i) = \omega_{A'} .$$

The pairs (m_i, n_i) are called by Igusa the numerical data of the resolution.

The numerical data can be used to compute the pull-back of the dualizing differentials on H. We have

*Partly supported by NSF grant MCS82-01159.

Lemma. $f^* \omega_H = \omega_{H'}(\Sigma(m_i - n_i)E_i')$ where

$E_i' = H' \cap E_i$.

Proof - We will compute locally on A and A'. Let ω be a basic differential on A. Let $\sigma = 0$ be a local equation of H in A. Then ω_H is free on the basis $\mathrm{Res}_H(\frac{\omega}{\sigma})$. Thus

$g^* \mathrm{Res}_H(\frac{\omega}{\sigma}) = \mathrm{Res}_{H'}(\frac{g^* \omega}{g^* \sigma})$. Let $\sigma' = 0$ and $\pi_i = 0$ be the local equations of H' and E_i so that $g^* \omega = \sigma' \cdot \Pi \pi_i^{m_i}$ and $g^* \omega = \Pi \pi_i^{n_i} \omega'$ where ω' is a basic differential on A'.

Thus $g^* \mathrm{Res}_H(\frac{\omega}{\sigma}) = \mathrm{Res}_{H'}(\frac{\omega'}{\sigma'} \Pi \pi_i^{n_i - m_i})$

$= \Pi \pi_i |_{H'}^{n_i - m_i} \mathrm{Res}_{H'}(\frac{\omega'}{\sigma'})$ where

the last residue is a basic differential of $\omega_{H'}$. This equation proves the lemma. Q.E.D.

Returning to rational singularities we have

Proposition (Char. 0) The hypersurface H has rational singularities if and only if the numerical data satisfy $m_i \leq n_i$ for all i.

Proof - The hypersurface H is always Cohen-Macaulay. Thus to see when H has rational singularities we need to check when the dualizing differentials ω_H pull-back to regular differentials on H'. By the lemma this happens if and only if $m_i \geq n_i$ for all i. Q.E.D.

The simplest example of a singular hypersurface is a cone over a smooth base. Let $H = (h=0)$ be the zeroes of a homogeneous polynomial h of degree d on $\mathbb{A}^n = A$. Assume that the projective divisor $h=0$ in $\mathbb{P}^{n-1}$ is smooth. If A' is the blowup of the origin 0 in $\mathbb{A}^n$, then $g^{-1}(H) = H' + dE$ where H' and E are smooth divisors meeting transversally. By an easy computation $g^* \omega_A((n-1)E) = \omega_{A'}$. Thus our cone H has rational singularities if and only if the degree $d <$ the number of variables n.

There is an equivalent definition of rational singularities. The other definition is more the reason for being of the concept.

Theorem (Char. 0) Let $f : X \to Y$ be a proper birational morphism with X a smooth variety. Then Y has rational singularities if and only if Y is normal (i.e., $f_* \mathcal{O}_X = \mathcal{O}_Y$) and $R^i f_* \mathcal{O}_X = 0$ if $i > 0$.

This equivalence is proved in [4]. The proof uses duality and the vanishing theorem ($R^i f_* \omega_X = 0$ for $i > 0$) of Granert-Riemannschneider.

Recently J. Köllar [5] has proved better vanishing theorems and applied his ideas to rational singularities to get an improvement of this theorem. His result is

Theorem (Char. 0) Let $f : X \to Y$ be a surjective morphism between projective varieties with X smooth. Let F be the geometric general fiber of f. Assume that F is connected. Then the following are equivalent

i) Y is normal and $R^i f_* \mathcal{O}_X = 0$ for all $i > 0$, and

ii) Z has rational singularities and $H^i(F, \mathcal{O}_F) = 0$ for $i > 0$.

This result implies a more general result where f is a proper surjective morphism and X has rational singularities. In this case the fiber F has rational singularities.

Köllar result may be applied to improve my results on collapsing homogeneous bundles [3]. Let $\mathbb{V}$ be a representation of a connected algebraic group G. Let F be a closed subvariety of $\mathbb{V}$ which is stabilized by a parabolic subgroup P of G. Then $G \cdot F$ is a closed subset of $\mathbb{V}$ which is in fact the image of the projective morphism $f : G \times_P F \longrightarrow \mathbb{V}$. My result now is

Theorem (Char. 0) If the action of P on F is completely reducible and F has rational singularities, then $G \cdot F$ has rational singularities.

The assumption implies the vanishing of the higher direct image of the structure sheaf of $G \times_P F$ by [3]. Thus the generalization of Köllar result proves this theorem. Previously I could only show that $G \cdot F$ was normal and Cohen-Macaulay.

Another recent development is worth noting. These results are valid in any characteristic. In fact the proof uses the Forbeneous in characteristic p. Recall that if B is a Borel subgroup of a connected subgroup G, then the closure of a B-orbit in the homogeneous space G/B is called a Schubert variety. The main result is

Theorem Schubert variety have rational singularities and also the cones over Schubert varieties which are induced from embeddings of G/B via complete linear systems.

The last development about this are contained in [6] [7] [8] and [1]. Many special cases were of course well-known.

References

[1] H.H. Andersen, preprint.

[2] G. Kempf, On the geometry of a theorem of Riemann, Annals of Math. 98(1973), 178-185.

[3] ________, On the Collapsing of Homogeneous Bundles, Invent. Math. 37(1976), 229-239.

[4] G. Kempf, F. Knudsen, D. Mumford, B. Saint-Donat, Toroidal Embeddings I, Lecture note in Math. 339, Springer-Verlag, New York 1973.

[5] J. Kollar, Higher Direct Images of Dualizing Sheaves

[6] V.B. Mehta and A. Ramanathan, Frobenius splitting and cohomology vanishing for Schubert varieties, Math. Ann.

[7] S. Ramanan and A. Ramanathan, Projective normality of flag varieties and Schubert varieties, Invent. Math.

[8] A. Ramanathan, Schubert Varieties are Arithmetically Cohen-Macaulay

Department of Mathematics
The Johns Hopkins University
Baltimore, MD 21218
U.S.A.

Contemporary Mathematics
Volume 58, Part I, 1986

Remarks on moduli spaces of complete intersections

A. S. Libgober and J. W. Wood *

1. Introduction.

In this paper we give an algorithm to compute the dimensions of the space of moduli of a complete intersection. This extends the treatment of hypersurfaces given by Kodaira and Spencer [8, §14]. One consequence, explained in §4, is that we can find diffeomorphic complete intersections of dimension three (and homeomorphic ones of dimension two) which lie in different dimensional components of the moduli space.

Our starting point is the results of E. Sernesi who showed [14] that the family $\mathcal{V}$ of complete intersections of fixed multidegree d whose defining polynomials have coefficients close to those of a given variety V_n, $n \geq 2$, is a complete complex analytic family of deformations of V except in the case of K-3 surfaces: n = 2 and d = (4), (3,2), or (2,2,2). It follows that any sufficiently small deformation of V is again a global complete intersection of the same multidegree. We let

1980 *Mathematics Subject Classification*. Primary 14J15, 14M10, 32G05, 32G13.
* The authors were partially supported by the National Science Foundation.

$V_n(d_1,\ldots,d_r)$, or simply V_n, denote a complete intersection of codimension r in P_{n+r} with multidegree $d = (d_1,\ldots,d_r)$ and $d_i \geq 2$.

Proposition 1 (Sernesi). $H^1(V_n, T_{P_{n+r}}|V_n) = 0$ for $n \geq 2$ except for the K-3 surfaces.

Sernesi used this in proving that the space of infinitesimal deformations of V is all of $H^1(V,T_V)$. It follows that the dimension, denoted m(V), of any effective, complete family of deformations is dim $H^1(V,T_V)$, [8, §§6 and 11].

To compute $H^1(V,T_V)$ we use the exact cohomology sequence corresponding to the exact sequence of sheaves [3, p.182]

(1) $$0 \longrightarrow T_{V_n} \longrightarrow T_{P_{n+r}}|V \longrightarrow N \longrightarrow 0$$

where N is the normal sheaf of V_n in P_{n+r}. In §3 we prove

Proposition 2. For any complete intersection except for the quadratic hypersurface

$$H^0(V,T_V) = 0 \qquad d \neq (2).$$

In general if V is a compact complex analytic manifold and $H^0(V,T_V) = 0$, the Kuranishi space is a local space of moduli for V. In our case, by Sernesi's result, its dimension is m(V) = dim $H^1(V,T_V)$, cf. [16, Chapter 2], [17, Theorem 1.1]. It also follows from $H^0(V,T_V) = 0$ that V admits no continuous group of analytic automorphisms. Under the stronger assumption that V has ample canonical bundle, Kobayashi showed [9] that Aut V is finite and Narasimhan and Simha [12] showed the existence of a global moduli space. They show the set of isomorphism classes of complex structures on V with ample canonical bundle has a natural structure as a Hausdorff complex space given in a neighborhood of a particular structure V_t by the Kuranishi space of V_t modulo Aut V_t.

As a consequence of Proposition 2 we can extend Kobayashi's result in the case of complete intersections.

Corollary. If V is a complete intersection of dimension ≥ 2 and not a K-3 surface or a quadratic hypersurface, then Aut V is finite.

Proof. Let $G = \text{Aut } P_{n+r} = PGL(n+r+1)$. The stabilizer $N_G(V) = \{ g \in G \mid gV \subset V\}$ is an algebraic group, cf. [1, p.97]. The proof of Theorem 8.2 in [10, I p.479] shows that except for K-3 surfaces, Aut V = $N_G(V)$. But Proposition 2 implies that Aut V is zero dimensional for degree > 2. The Corollary follows.

For a K-3 surface Aut V may be infinite; in [15, p.288] Severi gave an example of a surface of degree 4 with infinite automorphism group. In [11] Matsumura and Monsky give a more algebraic proof, which holds also in nonzero characteristic, that Aut V is finite for hypersurfaces (with the same exceptions.) They also prove that for a generic hypersurface V, Aut V = 0.

In §2 we present a formula for m(V). The proof of Proposition 2 is given in §3. Finally §4 contains examples in dimensions 2 and 3 of complete intersections whose moduli spaces have components of different dimensions and a further survey of results on components.

2. Computing m(V).

We will need the following facts about the cohomology of line bundles on V. For $d = (d_1,\dots,d_r)$ and for any $s \le r$ define

$$q(\ell,n,s,d) = \dim H^0(V_n(d_1,\dots,d_s),O_V(\ell)).$$

Lemma 1. Let V_n be a complete intersection of multidegree $d = (d_1,\dots,d_r)$ with $d_i \ge 2$.

(a) $H^i(V,O(\ell)) = 0$ for $i \ne 0,n$.

(b) The function q is determined by the recurrence relation

$$q(\ell,n,r,d) = \begin{cases} 0 & \ell < 0 \\ \binom{n+\ell}{\ell} & r = 0 \\ q(\ell,n+1,r-1,d) - q(\ell-d_r,n+1,r-1,d) & r > 0 \end{cases}$$

(c) $$\Sigma_{\ell=0}^{\infty} q(\ell,n,r,d)t^\ell = (1-t)^{-n-r-1}\,\Pi_{i=1}^{r}(1-t^{d_i})$$

Proof. (a) is found for example as an exercise in [3, p.231]. (c) is proved in [13, p.131]. (b) follows from (c) and conversely. Alternatively (a) and (b) follow by induction on r from the case of projective space [3, p.225] using the exact cohomology sequence coming from the sequence [4, 16.2.1]

$$0 \longrightarrow O_X(\ell-d_r) \longrightarrow O_X(\ell) \longrightarrow O_{X\cap W}(\ell) \longrightarrow 0$$

where $X = V_{n+1}(d_1,\dots,d_{r-1})$ and $W = V_{n+r-1}(d_r)$.

We have emphasized the recurrence relation (b) because it is useful for computation. There is another interpretation of the function q which may help to clarify its behavior. Writing the power series (c) as

$$(1+t+t^2+\dots)^{n+1}\,\Pi_{i=1}^{r}(1+t+\dots+t^{d_i-1}),$$

we see that the coefficient of t^ℓ is equal to the number of nonnegative integer solutions to the equation

$x_1 + \dots + x_{n+r+1} = \ell$ with $0 \le x_i < d_i$ for $i = 1,\dots,r$.

This of course equals the number of integer points inside an obviously defined polyhedron.

The function $q(\ell,n,r,d)$ is symmetric in $(d_1,\dots,d_r)$ but it is not polynomial and hence does not depend only on the elementary symmetric polynomials. On the other hand if

$$\ell > \Sigma(d_i - 1) - n - 1,$$

then by the Kodaira vanishing theorem [4, 18.2.2]

$$q(\ell,n,r,d) = \chi(V,\mathcal{O}(\ell))$$

which by the Riemann-Roch formula is a polynomial in $\ell + \tfrac{1}{2}c_1(V)$ and the Pontryagin classes of V [4, p.150] and hence is polynomial in ℓ, the total degree, Πd_i, and the first n elementary symmetric functions of $(d_1,\dots,d_r)$.

Theorem. If V_n is a complete intersection of multidegree $(d_1,\dots,d_r)$ with $d_i \ge 2$ and with $n \ge 2$ and not a K-3 surface or a quadratic hypersurface, then the number of moduli

$$m(V) = 1 - (n + r + 1)^2 + \Sigma_{i=1}^{r} q(d_i,n,r,d).$$

Proof. We first find $\dim H^0(V,T_{P_{n+r}}|V)$. Tensoring the exact sequence [3, p.182]

$$0 \longrightarrow \mathcal{O}_{P_{n+r}} \longrightarrow \mathcal{O}_{P_{n+r}}(1)^{n+r+1} \longrightarrow T_{P_{n+r}} \longrightarrow 0$$

with $\mathcal{O}_V$, since T_P is flat we obtain

$$0 \longrightarrow \mathcal{O}_V \longrightarrow \mathcal{O}_V(1)^{n+r+1} \longrightarrow T_{P_{n+r}}|V \longrightarrow 0$$

The corresponding cohomology sequence gives

$$0 \longrightarrow H^0(V,\mathcal{O}_V) \longrightarrow H^0(V,\mathcal{O}_V(1))^{n+r+1} \longrightarrow$$
$$H^0(V,T_{P_{n+r}}|V) \longrightarrow 0$$

Because V is not contained in any proper linear subspace of P_{n+r} we have $\dim H^0(V,O_V(1)) = n + r + 1$, in agreement with Lemma 1. Hence $\dim H^0(V,T_{P_{n+r}}|V) = (n+r+1)^2 - 1$.

In view of Propositions 1 and 2 the cohomology sequence corresponding to the sequence (1) becomes

$$(2) \quad 0 \longrightarrow H^0(V,T_{P_{n+r}}|V) \longrightarrow H^0(V,N) \longrightarrow H^1(V,T_V) \longrightarrow 0$$

Since the normal sheaf N is isomorphic to

$$O_V(d_1) + \dots + O_V(d_r),$$

the theorem follows from Lemma 1.

A table of values of m(V) in cases of low degree is given at the end of §4.

For a hypersurface of degree d we have

$$q(d,n,1,d) = \binom{n+1+d}{d} - 1$$

by Lemma 1 and so we obtain the formula

$$m(V) = \binom{n+d+1}{d} - (n+2)^2 \quad \text{for } d > 2$$

of Kodaira and Spencer [8, (14.10)].

A quadratic hypersurface V is rigid, $H^1(V,T_V) = 0$ [8, p.406], and our formula gives instead

$$\dim H^0(V,T_V) = \tfrac{1}{2}(n+1)(n+2).$$

For the K-3 surfaces the formula gives 19, the dimension of the image of δ^* in (2).

As the codimension r increases a closed form expression for m(V) becomes more complicated. For example if r = 2

$$m(V) = -1 - (n+3)^2 + \binom{n+2+d_1}{d_1} + \binom{n+2+d_2}{d_2} - \binom{n+2+d_2-d_1}{d_2-d_1} - \delta_{d_2}^{d_1}$$

assuming $d_2 \geq d_1$. Here δ is the Kronecker function.

3. A vanishing lemma.

In this section we compute $H^0(V,T_V)$. By Serre duality this group is isomorphic to $H^n(V,\Omega_V{}^1(\Sigma(d_i-1) - n - 1))$. If V has ample canonical bundle, $\Sigma(d_i-1) - n - 1 > 0$ and this group vanishes by a criterion of Akizuki and Nakano [8, (11.12)]. We reduce all but one of the remaining cases to their result using a theorem of Bott and an idea of Kodaira and Spencer [8, Lemma 14.2].

Lemma 2. Let V_n be a complete intersection. Then

$$H^q(V,\Omega_V{}^p(k)) = 0 \text{ for } p+q \geq n+1 \text{ and } k > p-q.$$

Proof. For $k > 0$ this is the result of Akizuki and Nakano. The proof is by induction on the codimension r of V. For $r = 0$, $V = P_n$, and our lemma follows from Bott's result [8, p.405]. Note that we may assume $q \geq 1$ since $\Omega^{n+1} = 0$.

For the inductive step assume V_n is a hypersurface of degree j in the complete intersection W_{n+1} of codimension r-1. The pair of exact sequences

$$0 \longrightarrow \Omega''_W{}^p \longrightarrow \Omega_W{}^p \longrightarrow \Omega_V{}^p \longrightarrow 0$$
$$0 \longrightarrow \Omega_W{}^p \longrightarrow \Omega''_W{}^p(j) \longrightarrow \Omega_V{}^{p-1} \longrightarrow 0$$

of Kodaira and Spencer [7, (3) and (4)] tensored with $\mathcal{O}_W(k+j)$ and $\mathcal{O}_W(k)$ yield a pair of exact sequences from whose cohomology sequences we take two short sections:

$$H^{q-1}(V,\Omega_V{}^{p+1}(k+j)) \longrightarrow H^q(W,\Omega''_W{}^{p+1}(k+j)) \longrightarrow H^q(W,\Omega_W{}^{p+1}(k+j))$$
$$H^q(W,\Omega''_W{}^{p+1}(k+j)) \longrightarrow H^q(V,\Omega_V{}^p(k)) \longrightarrow H^{q+1}(W,\Omega_W{}^{p+1}(k)).$$

Since $\dim W = n+1$ and $j \geq 2$, the hypotheses are satisfied by the first and last groups in the first sequence and the last group in the second sequence. Since W has lower codimension, the last groups are zero by induction. For large k (so $k+j > 0$) the first group in the first sequence is zero by the criterion of Akizuki and Nakano. The lemma follows by induction on $-k$.

Our lemma implies Proposition 2 provided

$$\Sigma(d_i-1) - n - 1 > 1 - n$$

hence for all cases except $d = (2)$, (3) and, $(2,2)$. The case $d = (3)$ uses a more complete form of Bott's theorem with $W = P_{n+1}$. It is covered by [8, Lemma 14.2].

In the case of a complete intersection of two quadrics the vanishing results above are not sufficient. One can use the following alternative argument. We have a commutative diagram

$$\begin{array}{ccccccccc} & & & & & & 0 & & \\ & & & & & & \big| & & \\ & & & & & & H^0(V,T_V) & & \\ & & & & & & \big| & & \\ 0 & \to & H^0(V,O_V) & \to & H^0(V,O_V(1))^{n+r+1} & \to & H^0(V,T_{P_{n+r}}|V) & \to & 0 \\ & & & & \mu\big| & & \big| & & \\ & & & & H^0(V,O_V(d_1) + \dots + O_V(d_r)) & = & H^0(V,N) & & \end{array}$$

The rightmost 0 comes from H&S1.$(V,O_V) = 0$ for $n \geq 2$ and the map μ is given by

$$(R_0,\dots,R_{n+r}) \longrightarrow \left(\Sigma_{i=0}^{n+r} R_i \frac{\partial F_1}{\partial x_i},\dots, \Sigma_{i=0}^{n+r} R_i \frac{\partial F_r}{\partial x_i}\right)$$

where

$R_i = \Sigma_{i=0}^{n+r} r_{ij}x_j$ are linear forms and $F_1 = 0,\dots, F_r = 0$ are the defining equations of $V_n(d_1,\dots,d_r)$. The kernel of μ is determined by the conditions

$$\Sigma_{i=0}^{n+r} R_i \frac{\partial F_k}{\partial x_i} = \Sigma_{l=1}^{r} a_{kl}F_l \quad \text{for} \quad k = 1,\dots,r.$$

For $d = (2,2)$ we take

$$F_1 = x_0^2 + \dots + x_{n+2}^2$$
$$F_2 = c_0 x_0^2 + \dots + c_{n+2} x_{n+2}^2$$

with $c_i \neq c_j$ for $i \neq j$. The equations imply $r_{ij} = 0$ for $i \neq j$ and $r_{ii} = \frac{1}{2} a_{11}$ for $i = 0,\dots,n+2$. Hence $\dim \ker\mu = 1$. This implies $H^0(V,T_V) = 0$. This completes the proof of Proposition 2.

In the case of a quadric hypersurface the computation of $\ker\mu$ yields

$$\dim H^0(V,T_V) = \tfrac{1}{2}(n+1)(n+2)$$

in agreement with the remark in §2.

4. Examples and remarks.

In [5, 6] Horikawa gave examples which show that the moduli space of algebraic structures on a given smooth 4-manifold can have different components of different dimensions. In [10] the authors showed that in any complex dimension > 2 and for any positive integer k, there are k distinct complete intersections which are all diffeomorphic. In fact they have equivalent underlying almost complex structures. However they lie in distinct irreducible components of the moduli space of complex structures on the underlying smooth manifold. We obtained the same result in dimension two for structures on a homeomorphism type. Recent work of Catanese [2] provides for any k a homeomorphism type of real dimension 4 supporting complex structures lying in components of the moduli space of k different dimensions. It is natural to expect similar behavior among complete intersections.

For surfaces, the homotopy type is determined by the middle Betti number and the signature and type of the intersection pairing which can be computed from the total degree and the first two symmetric functions of d. By the work of Mike Freedman the homeomorphism type of these manifolds is determined by the homotopy type. We find the varieties $V_2(6,6,6,2,2,2,2)$ and $V_2(8,4,4,3,3,3)$ are homeomorphic but have $m(V) = 7509$ and 9546 respectively.

There is a way to generate larger sets of homeomorphic surfaces from pairs. Suppose $d = (d_1,\ldots,d_r)$ and $e = (e_1,\ldots,e_s)$ are two multidegrees with the same total degree and symmetric functions σ_1 and σ_2. Let $de = (d_1,\ldots,d_r,e_1,\ldots,e_s)$. Then the three multidegrees dd, de, and ee also have the same invariants σ_1, σ_2, and total degree. Similarly for ddd, dde, dee, and eee, etc. It is reasonable to expect that the corresponding m(V)'s will all be different. Unfortunately we have no general way to prove this.

In case $d = (10,10,4,3,3)$ and $e = (12,6,5,5,2)$ which give homeomorphic surfaces we find

multidegree	$m(V_2)$
d	23356
e	27005
dd	1695364
de	2234720
ee	2758226

Computation of the four cases which come from juxtaposing three terms and of the five cases which come from juxtaposing four terms gives four and five distinct moduli dimensions respectively.

For 3-folds the total degree, σ_1, σ_2, σ_3 determine the diffeomorphism type [10, I p.480]. In this case the varieties $V_3(14,14,5,4,4,4)$ and $V_3(16,10,7,7,2,2,2)$ are diffeomorphic with moduli dimensions 1028748 and 1191130 respectively. As above larger sets of diffeomorphic varieties can be generated. We have checked that the moduli dimensions are all different through the case of five distinct but diffeomorphic complete intersections.

These computations and the ones for the small table below were done with the aid of a computer using the recursive definition of the function q given in Lemma 1(b).

Table of m(V) for some complete intersections of low degree

multidegree	dim 2	dim 3
3	4	10
4	19 (K-3)	45
2,2	2	3
5	40	101
6	68	185
3,2	19 (K-3)	34
4,2	44	89
2,2,2	19 (K-3)	27
3,2,2	46	73

References.

[1] A. Borel, Linear algebraic groups, Benjamin, 1969.

[2] F. Catanese, On moduli spaces of surfaces of general type, J. Diff. Geometry 19 (1984) 483-515.

[3] R. Hartshorne, Algebraic Geometry, Springer-Verlag, 1977.

[4] F. Hirzebruch, Topological methods in algebraic geometry, Springer-Verlag, 1966.

[5] E. Horikawa, On deformations of quintic surfaces, Invent. Math. 31 (1975) 43-85.

[6] E. Horikawa, Algebraic surfaces of general type with small c_1^2 IV, Invent. Math. 50 (1979) 103-128.

[7] K. Kodaira and D. C. Spencer, On a theorem of Lefschetz and the lemma of Enriques-Severi-Zariski, Proc. Nat. Acad. Sci. 39 (1953) 1273-1278.

[8] K. Kodaira and D. C. Spencer, On deformations of complex analytic structures, I,II, Ann. of Math. 67 (1958) 328-466.

[9] S. Kobayashi, On the automorphism group of a certain class of algebraic manifolds, Tôhoku Math. J. 11 (1959) 184-190.

[10] A. Libgober and J. W. Wood, Differentiable structures on complete intersections I, Topology 21 (1982) 469-482 and II, Proc. Symp. Pure Math. 40 (1983) Part 2, 123-133.

[11] H. Matsumura and P. Monsky, On the automorphisms of hypersurfaces, J. Math. Kyoto Univ. 3 (1964) 347-361.

[12] M. S. Narasimhan and R. R. Simha, Manifolds with Ample canonical class, Invent. Math 5 (1968) 120-128.

[13] P. Orlik and P. Wagreich, Singularities of algebraic surfaces with C^* action, Math. Ann. 193 (1971) 121-135.

[14] E. Sernesi, Small deformations of global complete intersections, Bollettino della Unione Matematica Italiana 12 (1975) 138-146.

[15] F. Severi, Complementi alla teoria della base par la totalita della curve di una superficie algebrica, Rendiconti del Circ. Mat. di Palermo 30 (1910) 265-288.

[16] D. Sundararaman, Moduli, deformations and classifications of compact complex manifolds, Pitman, 1980.

[17] J. Wavrik, Obstructions to the existence of a space of moduli, Global Analysis, Princeton, 1969, pp. 403-414.

Department of Mathematics, Statistics, and Computer Science
University of Illinois at Chicago
Chicago, IL 60680

Contemporary Mathematics
Volume 58, Part I, 1986

THE SCHOTTKY GROUPS IN HIGHER DIMENSIONS

Madhav V. Nori

It seems likely that the analogues of Kleinian groups in dimension greater than one will provide a richer supply of compact complex manifolds. Here we shall generalise the classical Schottky groups to odd dimensions and thus provide compact complex manifolds of dimension ≥ 3 with a free group for fundamental group. During this talk, it was pointed out by N. Hitchin that such manifolds were already known to exist by a use of the twistor construction.

Let Γ be a finitely generated subgroup of $PGL_{n+1}(\mathbb{C}) = \text{Aut}\, \mathbb{P}^n(\mathbb{C})$ and let Ω be the (open) subset of those $x \in \mathbb{P}^n(\mathbb{C})$ which possess a neighborhood U such that $\{\gamma \in \Gamma \mid \gamma U \cap U \neq \emptyset\}$ is finite. What is the structure of the quotient of Ω by the action of Γ?

When $n = 1$, it is a theorem of L. Ahlfors that $\Gamma \setminus \Omega$ is a finite disjoint union of Riemann surfaces, each of which is the complement of a finite set in a compact Riemann surface.

When $n \geq 2$, however, $\Gamma \setminus \Omega$ has an unfortunate tendency to be non-Hausdorff. For example, let Γ be the cyclic group generated by the transformation γ, where

$$\gamma[z_0, z_1, \ldots, z_n] = [\lambda_0 z_0, \lambda_1 z_1, \ldots, \lambda_n z_n]$$

and assume that $0 < |\lambda_0| < |\lambda_1| < \cdots < |\lambda_n|$. It is then easy to see that Ω is precisely the complement in $\mathbb{P}^n(\mathbb{C})$ of the set of points with exactly one non-vanishing co-ordinate. If we let L_j and M_j be the linear subspaces of $\mathbb{P}^n(\mathbb{C})$ given by $z_0 = z_1 = \ldots = z_j = 0$ and $z_{j+1} = \ldots = z_n = 0$ respectively, then the quotient of $\mathbb{P}^n(\mathbb{C}) - (L_j \cup M_j)$ by Γ is a compact complex manifold M_j which is also open in $\Gamma \setminus \Omega$. Since Ω is connected, $\Gamma \setminus \Omega$ is also connected and therefore $\Gamma \setminus \Omega$ is not T_2.

QUESTION: With Γ finitely generated as above, consider the collection of

Γ-stable open subsets on which Γ acts properly discontinuously. Let Ω' be a maximal open subset in this collection. Can $\Gamma\backslash\Omega'$ be realized as the complement of a closed complex-analytic subset in a compact complex-analytic space M?

We shall now show that the classical Schottky groups have a generalisation to odd dimensions. Assume for the moment we are given the following data, given a natural number g:

(1) $2g$ open sets $R_1,\dots,R_g$, $S_1,\dots,S_g$ in $\mathbb{P}^{2k+1}(\mathbf{C})$ with the property that

(a) each of these open sets is the interior of its closure and

(b) the closures of the $2g$ open sets are pairwise disjoint, and

(2) elements $\gamma_i \in PGL_{2k+2}(\mathbf{C})$ such that $\gamma_i(R_i) = \mathbb{P}^{2k+1}(\mathbf{C}) - \bar{S}_i$.

Let Γ denote the subgroup of PGL generated by $\{\gamma_1,\gamma_2,\dots,\gamma_g\}$, let F be the complement of the union of all the R_i and the S_i, and finally put $\Omega' = \bigcup_{\gamma\in\Gamma}\gamma F$.

In the Schottky situation, the R_i and S_i are open discs in $\mathbb{P}^1(\mathbf{C})$. And exactly as in that situation one shows

(A) Γ is free on its generators $\{\gamma_1,\gamma_2,\dots,\gamma_g\}$,

(b) Ω' is open and $\Gamma\backslash\Omega'$ is precisely the quotient-space of F obtained by identifying the disjoint subsets ∂R_i and ∂S_i of F by the transformation γ_i , for $1 \le i \le g$.

We shall now construct the Schottky data (1) and (2) in $\mathbb{P}^{2k+1}(\mathbf{C})$. First choose $2g$ disjoint linear subspaces of dimension k in $\mathbb{P}^{2k+1}(\mathbb{C})$ to be denoted by $L_1,L_2,\dots,L_g$ and $M_1,M_2,\dots,M_g$. Now fix a natural number i such that $1 \le i \le g$ and choose a basis so that L_i and M_i are given by

$$z_0 = z_1 = \cdots = z_k = 0 \qquad \text{and} \qquad z_{k+1} = \cdots = z_{2k+1} = 0$$

respectively. Define $\phi_i\colon \mathbb{P}^{2k+1}(\mathbb{C}) \to \mathbb{R}$ by

$$\phi_i[z_0,\dots,z_{2k+1}] = \sum_{j=0}^{k} |z_j|^2 \Big/ \sum_{j=0}^{2k+1} |z_j|^2 .$$

Define R_i and S_i by

$$R_i = \{ v \in \mathbb{P}^{2k+1}(\mathbf{C}) : \phi_i(v) < \alpha\} \quad \text{and} \quad S_i = \{v \in \mathbb{P}^{2k+1}(\mathbf{C}) : \phi_i(v) > 1 - \alpha\} ,$$

where α is some positive real number. Define γ_i by

$$\gamma_i[z_0,\ldots,z_{2k+1}] = [\lambda z_0,\ldots,\lambda z_k, z_{k+1},\ldots,z_{2k+1}] ,$$

where $\lambda \in \mathbb{C}$ and $|\lambda| = \alpha/1 - \alpha$. Taking α very small we see that the R_i and S_i are contained in arbitrary neighborhoods of L_i and M_i respectively and their closures are therefore pairwise disjoint. And the γ_i have been constructed so that (2) holds.

Now assume $k \geq 1$. Thus the <u>real</u> codimension of L_i is $2k + 2 \geq 4$ and since R_i is a tubular neighborhood of L_i, it follows that F is simply connected. From (B) above it follows that the fundamental group of the compact complex manifold $\Gamma \backslash \Omega'$ is free on g generators.

We now include here the direct construction of compact complex manifolds with free fundamental group as pointed out to us by N. Hitchin. The point is that the collection of compact oriented manifolds possessing a conformally flat structure is closed under products and connected sums. In particular, if A_r denotes a compact oriented surface of genus r, then $A_r \times A_s$ carries a conformally flat structure, and so does $S^1 \times S^3$. Taking connected sums, we get compact oriented 4-manifolds X with a conformally flat structure and $\pi_1(X) \simeq G_1 * G_2 * \cdots * G_m$ where each G_i is isomorphic to $\mathbb{Z}$ or to $\pi_1(A_r) \times \pi_1(A_s)$ for arbitrary $r,s \geq 0$. The structure group of the tangent-bundle of X is reduced to $SO(4)$ once a Riemannian metric (compactible with the conformally flat structure) has been chosen and if M denotes the fibre-space on X associated to the principal homogeneous space $SO(4)/U(2)$, then it is known that M is a complex manifolds. And since $M \to X$ is a $\mathbb{P}^1(\mathbf{C})$-fibration, $\pi_1(M) \to \pi_1(X)$ is an isomorphism.

REFERENCE

B. MASKIT, "A characterisation of Schottky groups", Journal D'Analyse (1967), vol. 19, pages 227-230.

School of Mathematics
T.I.F.R.
Bombay 400005
INDIA

Contemporary Mathematics
Volume 58, Part I, 1986

COHOMOLOGY ON COMPLEX HOMOGENEOUS MANIFOLDS WITH COMPACT SUBVARIETIES [1]

by C. Patton and H. Rossi [2]

Abstract: *Let C^{p+q} be complex $(p+q)$-space endowed with a non-degenerate hermitian form h of signature (p,q). For $r \leq p$, $s \leq q$, let $M_{r,s}$ represent the space of $(r+s)$-dimensional subspaces of C^{p+q} on which h has signature (r,s). Then $SU(p,q)$ acts transitively on $M_{r,s}$. Let $V \to M_{r,s}$ be a homogeneous vector bundle, and $H^d(M_{r,s}, V)$ the associated cohomology groups. There is then a representation $\rho(d,V)$ of $SU(p,q)$ on this cohomology. For appropriate choices of d, V, $\rho(d,V)$ is realized as a subrepresentation of a tensor product of holomorphic and antiholomorphic series representations.*

I. Introduction

Let X be a complex manifold of dimension n. We suppose that X has these properties:

(1.1) There is an integer d such that every compact subvariety of X has dimension at most d;

(1.2) Every point of X lies on a compact subvariety of dimension d.

The technique for studying analytic objects on such a manifold is to remove them to the *space* $\mathcal{N}_d(X)$ *of d-dimensional connected compact subvarieties of* X.

Example 1. Suppose X is holomorphically convex; that is, for every discrete sequence $\{x_n\} \in X$, there is a holomorphic function f such that $\lim |f(x_n)| = \infty$. Then there is an integer d such that, but for an *exceptional* compact subvariety of dimension greater than d, (1.1) and (1.2) are satisfied. In fact, let E be this equivalence relation on X:

(1.3) $x \; E \; y$ if and only if x and y lie on the same conected compact subvariety of X.

Then X/E is the space $\mathcal{N}_d(X)$ as in the following theorem:

1.4. Theorem. *Suppose that X is holomorphically convex and E is the equivalence relation defined by (1.3). Then X/E is a Stein analytic space, and the quotient map $q : X \to X/E$ is proper. If $\mathcal{S} \to X$ is a coherent analytic sheaf on X, we have*

$$H^d(X, \mathcal{S}) \cong H^0(X/E, R^d q_*(\mathcal{S})),$$

[1] **1980 Mathematics Subject Classification 22E50.**

[2] **The second author was partially supported by the National Science Foundation under grant DMS 8401753.**

where R^dq_* is the direct image map.

Example 2. Suppose that X is a d-complete n-dimensional manifold. That is, there is a $\mathcal{C}^\infty$ real-valued function f defined on X such that

(a) for all k, $\{x \in X; f(x) \le k\}$ is compact,

(b) at every point $x \in X$, in local coordinates $z_1, \ldots, z_n$, the complex hessian of f:

$$\frac{\partial^2}{\partial z_i \partial \bar{z}_j} f(x)$$

has at least $n - d$ positive eigenvalues.

Then (1.1) holds, and further [1], if $\mathcal{S} \to X$ is any coherent analytic sheaf, $H^d(X, \mathcal{S})$ is infinite dimensional. Furthermore , evaluation of a class $\omega \in H^d(X, \mathcal{O})$ on a d-dimensional compact subvariety K defines a function $\hat{\omega}(K)$ on $\mathcal{N}_d(X)$. Of course condition (1.2) may fail, but if it also holds, we have this theorem of Andreotti and Norguet (there is an additional technical assumption on the hessian of f):

1.5. Theorem [1]. *With these hypotheses, $\mathcal{N}_d(X)$ has a unique structure of a Stein complex analytic space making the functions $\hat{\omega}$ holomorphic.*

In the case of Theorem 1.4, $X \to \mathcal{N}_d(X)$ is a fibration, but in the case of Theorem 1.5 in general it is not. The following example illustrates this.

Example 3. (The Penrose Correspondence). Let C^4 be endowed with the hermitian form h of signature (2,2):

$$h(z, z) = |z_1|^2 + |z_2|^2 - z_3|^2 - |z_4|^2.$$

Let M_1 (M_2) be the collection of complex lines (planes) in C^4 on which h is positive definite. Then M_1 satisfies the condition of theorem 1.5 with $d = 1$, and M_2 is the Stein space $\mathcal{N}_1(M_1)$. For $P \in M_2$, its corresponding compact subvariety is the set of all complex lines in P. Since any positive definite line lies in many positive definite planes, this is not a fibration. (This was in fact the example motivating the theorem of Andreotti and Norguet).

Penrose went much further with this. Let $L \to M_1$ be a line bundle whose restriction to every compact subvariety K is negative in the sense of Kodaira. Then $H^1(K, L) \neq 0$, and as K varies these groups define a vector bundle $\Theta(L)$ over M_2, and every class $\omega \in H^1(M_1, L)$ defines a section $\hat{\omega}$ of $\Theta(L)$. This is the Penrose correspondence:

(1.6) $$P : H^1(M_1, L) \to H^0(M_2, \Theta(L)).$$

The point of the Penrose correspondence is that the context on the left is physically significant, while that on the right is mathematically well-understood.

(1.6) can be made more precise by introducing the flag manifold

$$F = \{(V, W); V \in M_1, W \in M_2; V \subset W\}.$$

Denoting the two projections by μ and ν, this gives us the diagram

$$\begin{array}{ccccc} & & F & & \\ & \swarrow \mu & & \searrow \nu & \\ M_1 & & & & M_2 \end{array} \tag{1.7}$$

For L a line bundle over M_1, its lift to F is denoted $\mu^*(L)$. Then $\Theta(L)$ is the direct image sheaf $R^1\nu_*(\mu^*(L))$, and (1.6) is precisely the map $R^1\nu_*\mu^*$ defined on sections on M_1. In [4] it is shown that (for $SU(2,2)$-homogeneous bundles) this correpsondence is injective and the image lies in the holomorphic series of representations of $SU(2,2)$.

The results of example 3 have been generalized to appropriate homogeneous spaces for the groups $SU(p,q)$ in [2,3,6,8]. In [9], Rawnsley, Schmid and Wolf succeeded in extending these results to a very general context, which we shall now describe.

Let G be a semisimple Lie group with maximal compact subgroup K so that G/K is a hermitian symmetric space. Let H be another subgroup of G such that G/H also carries a complex structure. The case in which H is compact (and thus $H \subset K$), was studied by Schmid in [10], and further developed by Wells and Wolf in [11]. In the paper of Rawnsley,Schmid and Wolf, a spectral sequence (which already occurs in [4] and [11]) is constructed which effectively computes $H^d(G/H, L)$ for homogeneous line bundles L. However, the full strength of the Penrose correspondence generalizes only when an additional condition (of holomorphic compatability, (1.7) in [9]) is satisfied: it is required that $G/H \cap K$ carries a complex structure so that the natural maps μ,ν in this diagram

$$\begin{array}{ccccc} & & G/H \cap K & & \\ & \swarrow \mu & & \searrow \nu & \\ G/H & & & & G/K \end{array} \tag{1.7}$$

are holomorphic. In this case, the Penrose correspondence is defined as $P = R^d\nu_*\mu^*$ and the spectral sequence tells us precisely when P is injective. The image of P always lies in the holomorphic discrete series.

In this paper we shall show that the Penrose correspondence works as it does when (1.7) holds, becuase G/K is precisely the space $\mathcal{N}_d(G/H)$, and the ideas of Andreotti and Norguet apply. We shall show that when (1.7) fails, we do not map into the holomorphic series, and this is because $\mathcal{N}_d(G/H)$ is much larger. This was made clear by Wells and Wolf for compact H. We shall make these results explicit for the special case $G = SU(p,q)$.

II. Kodaira-Namba Space

First, let us give the precise definition of the space $\mathcal{N}_d(X)$.

2.1. Definition. *Let X be a complex manifold of dimension n. A* **family of compact connected submanifolds** *of X consists in (1) a fibration $\nu : F \to B$ of complex spaces,* *whose fibers are compact and connected manifolds, (2) a holomorphic map $\mu : F \to X$ such that μ is an injection on each fiber.*

2.2. Definition. *Let*

$$\begin{array}{ccc} F & \xrightarrow{\delta} & B \\ \nu\downarrow & & \\ X & & \end{array}$$

be a family of submanifolds of X and $f : P \to B$ a holomorphic map. The **induced family** *$f^*(\nu) : f^*(F) \to P$ is defined as follows:*

$$f^*(F) = \{(p, x) \in P \times F; f(p) = \nu(x)\}.$$

If $\pi_1 : f^*(F) \to P, \pi_2 : f^*(F) \to F$ are the natural projections, we define

$$f^*(\nu) = \pi_1, f^*(\mu) = \mu \circ \pi_2.$$

A family $\nu : F \to B$ is **universal** if every other family is induced by a unique holomorphic map to B. The main theorem of Kodaira and Namba [5,7]is that the set of compact connected submanifolds of X can be made into a universal family:

2.3. Theorem. *Let M be the set of compact connected submanifolds of X, and F the* **tautological bundle** *over M :*

$$F = \{(m, x); m \in M, x \in m\}.$$

Then F, together with the projections $\nu : F \to M, \mu : F \to X$ can be given complex structures so that they describe the universal family of submanifolds. If

$$\begin{array}{ccc} F & \xrightarrow{\beta} & B \\ \alpha\downarrow & & \\ X & & \end{array}$$

is another family, the inducing map $f : B \to M$ is defined by $f(b) = \alpha(\beta^{-1}(b))$.

If M is a compact conected submanifold of X, we define its normal bundle by the exact bundle sequence

$$0 \to T_M \to T_X \to N \to 0 \tag{2.4}$$

where T_M, T_X are the tangent bundles of M and X respectively. The following theorem comes out of the proof of Theorem 2.3.

2.5. Theorem. *The tangential dimension of M at K is $H^0(K, N)$. M is regular at K if and only if this number is constant under all small deformations of K; in particular this is assured if $H^1(K, N) = 0$.*

The case where $H^1(K, N) = 0$ was done by Kodaira in [5], where he introduced the necessary machinery. The theorem was completed by Namba in [7]. Now, if we are given a particular M in X we shall be interested only in the component of M containing M; this we shall denote by KN and call it the Kodaira-Namba space of M in X.

We shall apply these ideas in the context of [9] and compute the dimensions in theorem 2.5.

Let G be a semisimple Lie group, and H a closed subgroup. If $X = G/H$ carries the structure of a complex manifold, we have the Lie algebra decomposition

$$g = h \oplus q$$

where q is an ideal, and an ad$H$-invariant splitting into Lie subalgebras

$$q^C = q^+ \oplus q^-, \quad q^- = \bar{q}^+$$

defining the complex structure. That is, the H-module q defines a vector bundle on X which is its tangent bundle. Then q^+ defines, by the same identification, the holomorphic tangent bundle to X giving the complex structure. We shall denote this bundle on X by $\mathcal{T}(q^+)$. Now every element of g^C (considered as a left-invariant vector field on G) defines a vector field on X which is holomorphic, since G acts by holomorphic transformations. We denote this map $\Phi : g^C \to H^0(X, \mathcal{T}(q^+))$. We note that Φ is injective, and that, if $ev : H^0(X, \mathcal{T}(q^+)) \to q^+$ is the evaluation at eH, then $ev \circ \Phi : g^C \to q^+$ is just the projection map.

If K is a maximal compact subgroup of G we have the Cartan decomposition

$$g = k \oplus p.$$

Let $L = H \cap K$. Then

$$k^C = \ell^C \oplus (k^C \cap q^+) \oplus (k^C \cap q^-).$$

Defining $k^+ = k^C \cap q^+, k^- = k^C \cap q^-$, this splitting puts a complex structure on $M = K/L$, making it a compact complex subvariety of $X = G/H$. Now, the holomorphic normal bundle sequence (2.4) along M in X becomes

$$0 \to \mathcal{T}(k^+) \to \mathcal{T}(q^+) \xrightarrow{\pi} \mathcal{T}(p^C \cap q^+) \to 0, \tag{2.6}$$

that is, $p^C \cap q^+$ induces the normal bundle along K/L as L-module. We shall need this lemma:

2.7. Lemma. *Let $V \to K/L$ be a homogeneous vector bundle. Then V_{eH} is an L-module, and $ev : H^0(K/L, V) \to V_{eH}$ gives a correspondence between K-submodules of $H^0(K/L, V)$ and H-submodules of V_{eH}.*

2.8. Corollary. *$H^0(K/L, N)$ is the image under $\pi \circ \Phi$ of the K-orbit of $p^C \cap q^+$.*

Proof. Since $ev \circ \pi \circ \Phi$ is the identity on $p^C \cap q^+$, by the lemma, $\pi \circ \Phi$ of its K-orbit must fill out all of $H^0(K/L, N)$. Now, in [9], they prove the following dichotomy:

(a) If the K-orbit of $p^C \cap q^+$ is a proper subspace p^+ of p^C, then, (for p^- the conjugate of p^+), we have the decompositions

$$g^C = k^C \oplus p^+ \oplus p^-$$

and

$$g^C = \ell^C \oplus (p^+ + q^+) \oplus (p^- + q^-).$$

These put complex structures on $G/K, G/L$ respectively, so that the maps in

$$(1.7) \qquad \begin{array}{ccc} & G/H \cap K & \\ \swarrow \mu & & \searrow \nu \\ G/H & & G/K \end{array}$$

are holomorphic (condition (1.7) holds).

(b) If $p^C \cap q^+$ generate p^C over K, no such choice of complex structure exists.

From this we get the following theorem:

2.9. Theorem.

(a) In case (1.7) holds,

$$H^0(K/L, N) \cong p^+, \quad H^0(K/L, T_{K/L}) \cong k^C \oplus p^-$$

and the G-orbit of K/L is open in KN space.

(b) In case (1.7) fails,

$$H^0(K/L, N) \cong p^C, \quad H^0(K/L, T_{K/L}) \cong k^C$$

and the G^C-orbit of K/L is open in KN space.

Proof. First of all, from the corollary we conclude that $H^0(K/L, q^+) \cong \Phi(g^C)$, i.e, that g^C is precisely the set of sections of T_X on K/L. Now, first we assume that we are in case (a). Again by the lemma, since p^+, p^- are ad$K$-invariant,

$$(2.10) \qquad \pi \circ \Phi(p^+) = H^0(K/L, N) \cong g^C / H^0(K/L, T_{K/L}),$$

and $\pi \circ \Phi(p^-) = 0$, so that $\Phi(p^-) \subset H^0(K/L, T_{K/L})$. Thus $\pi \circ \Phi(k^C \oplus p^-) = 0$. By (2.10), the kernel of $\pi \circ \Phi$ can be no larger, so $\pi \circ \Phi$ is an isomorphism on p^+, and the dimension of the KN space is that of p^+, which is the same as the complex dimension of G/K, so the latter is open.

(b) This time, by the corollary, we have

$$(2.11) \qquad \pi \circ \Phi(p^C) = H^0(K/L, N) = g^C / H^0(K/L, T_{K/L}).$$

Thus the kernel of $\pi \circ \Phi$ can be no larger than k^C, so that it is an isomorphism on p^C, and the theorem is proven.

In case H is compact, and thus $H \subset K$, this calculation shows that the deformation spaces introduced by Wells and Wolf in [11] coincide with the corresponding KN spaces.

2.12. Corollary. *If H is compact, then*
(a) if (1.7) holds, G/K is the KN space of K/H in G/H, and
(b) if (1.7) fails, KN space is

$$\{gK^C \in G^C/K^C; \quad g \cdot K/H \subset G/H\}. \tag{2.13}$$

Proof. (a). In this case, the projection map $G/H \to G/K$ is holomorphic, has compact fibers and G/K is a Stein domain. Then theorem 1.4 applies. (b). Let D be the domain in (2.13). For $m = gH$ in D let $K_m = gK/H$. Now suppose that m_n is a sequence in D converging to m_0 in KN. Then, in G/H, $K_{m_n} \to K_0$, so they all lie in a relatively compact subset of G/H. By lemma 2.5.1 of [11], we then also have m_0 in D. Thus D is closed in KN , so by the theorem coincides with KN.

Now, returning to the general case, let KN represent the Kodaira Namba space of K/L in G/H, $\nu : F \to KN$ the *tautological bundle* of theorem 2.3, and $\mu : F \to G/H$ as in that theorem. Let $V \to G/H$ be a homogeneous vector bundle, and $\mathcal{V}$ its sheaf of sections. Then since ν is a proper map, $R^p\nu_*\mu^*(\mathcal{V})$ is a coherent analytic sheaf on KN. Since we are in the homogeneous case, this is also locally free, and therefore the sheaf of sections of a vector bundle we denote as $\Theta(\mathcal{V})$. Let d be the dimension of K/L (and therefore also the dimension of the fiber of ν). We define the **Penrose transform** as the induced cohomology map

$$P = R^d\nu_*\mu^* : H^d(G/H, \mathcal{V}) \to \mathcal{H}'(\mathcal{K}\mathcal{N}, \times(\mathcal{V})). \tag{2.14}$$

It is easy to see that G acts hololmorphically on KN, and that P is an intertwining operator for G: P commutes with the G-action. In case (1.7) holds, the G-orbit of K/H is G/K, and is open in KN, so the image of P is in a holomorphic representation of G (the space of section of a holomorphic vector bundle on G/K). However if (1.7) fails, G/K, the G-orbit of K/H, is a real form of KN, and the image of P does not lie in a holomorphic representation, but in the tensor product of a holomorphic and an antiholomorphic representation of G.

III. SU(p,q)

$h(z, \bar{w})$ is the standard hermitian form of signature (p, q) on C^{p+q}:

$$h(z, \bar{w}) = \sum_{n=1}^{p} z_i\bar{w}_i - \sum_{n=p+1}^{p+q} z_i\bar{w}_i. \tag{3.1}$$

By C^p or C^q we shall mean the summands of the expression $C^{p+q} = C^p \oplus C^q$. $G(d, n)$ is the grassmanian of d-dimensional subspaces of C^n. $M_{r,s}$ is the domain in $G(r+s, p+q)$ of subspaces of signature (r, s). At the extremes we get M_+, M_-, the domains respectively of positive definite, negative definite lines in C^{p+q}; and M_p, M_q , the Seigel domains of positive definite p-planes, negative definite q-planes. We think of M_q as being the same as M_p, but carrying the conjugate complex structure (except of course when p or q is 0, in which case one of them reduces to a point). By $E(k)$ we mean the kth tensor power of the determinant of the tautological bundle on $G(d,n)$. By $U(p,q)$ we mean the group of linear transformations on C^{p+q} leaving the form h invariant, and $SU(p,q)$ are those of determinant one.

In Patton-Rossi [8] and Rawnsley-Schmid-Wolf [9] unitary representations are constructed on the spaces $M_{r,0}$, in cohomology of degree $r(p-r)$. The Penrose transform intertwines these with holomorphic representations on Seigel space. In Patton-Rossi [8], Mantini [6], and Blattner-Rawnsley [2] these are shown to be subrepresentations of tensor powers of the metaplectic representation of $SU(p,q)$. In all of this, the condition (1.7) of [9] holds, but fails as soon as *both* r and s are not zero. Here we shall now describe in detail in this case how the results of section 2 apply.

Here $G = SU(p,q), K = S(U(p)\times U(q)), H = S(U(r,p-r)\times U(s,q-s)), L = S(U(r)\times U(p-r)\times U(s)\times U(q-s)$, and $G/H = M_{r,s}$. The dimension of $M_{r,s}$ is $(r+s)[p+q-(r+s)]$. Let $(P,Q)\in M_p\times M_q$; this space is of dimension $2pq$. For such P,Q, let

$$M(P,Q) = G(r,P)\times G(s,Q) = \{L+L';\ L\in P \quad and \quad L'\in Q\}.$$

This space is of dimension $r(p-r)+s(q-s)$ and is embedded in $M_{r,s}$ by the map $\nu_{P,Q} : (L,L')\to L\oplus L'$. As such it is of maximal dimension among compact subvarieties of $M_{r,s}$. The codimension , and thus the rank of the normal bundle to $M(P,Q)$ is $r(q-s)+s(p-r)$. Now, $M_0 = M(C^p, C^q)$ is H/L, and the maps $\nu_{P,Q}$ clearly describe a holomorphic family of deformations of M_0 in $M_{r,s}$, based on $M_p\times M_q$. A quick check of dimensions verifies that $M_p\times M_q$ is a candidate for KN space by theorem 2.8; we now give a direct geometric proof, for which we acknowledge the help of Ron Donagi.

Let V be the direct sum of subspaces V_1 and V_2. Let

$$G = G(d,V), \quad G_i = G(d_i,V_i), \quad i=1,2$$

be the Grassmanians with $d_1+d_2 = d$. A point in $Gr(d,V)$ shall be denoted W, and by $\mathcal{W}$ we shall mean the tautological bundle (or its sheaf of sections) over $Gr(d,V)$. Now the map $G_1\times G_2\to G$, given by

$$(W_1,W_2)\to W_1+W_2$$

is an injection. Let M be the image. Let $\mathcal{T}_G$ be the tangent bundle to G. We have the exact sequence

$$0\to \mathcal{T}_M\to \mathcal{T}_G\to \mathcal{N}\to 0$$

defining the normal bundle $\mathcal{N}$.

3.2. Proposition. *$H^0(M,\mathcal{N}) = Hom(V_1,V_2)\oplus Hom(V_2,V_1)$ unless $d_i = dim V_i$ in which case the summand with V_i as target does not appear, or $d_i = 0$ in which case the summand with V_i as source does not appear.*

Proof. For $W = W_1\times W_2\in M$

$$\mathcal{T}_{G,W} = Hom(W, V/W) = Hom(W_1\oplus W_2, V/(W_1\oplus W_2)) =$$

$$= \begin{pmatrix} Hom(W_1, V_1/W_1) & Hom(W_1, V_2/W_2) \\ Hom(W_2, V_1/W_1) & Hom(W_2, V_2/W_2) \end{pmatrix}.$$

Since $\mathcal{T}_M$ consists of the matrices of the form

$$\begin{pmatrix} * & 0 \\ 0 & * \end{pmatrix},$$

$\mathcal{N}$ consists of those whose entries are on the off-diagonal.

Let h be in $H^0(M, \mathcal{N})$, $h = h_1 + h_2$. Then h_1 is in $H^0(M, Hom(\mathcal{W}_1, \mathcal{V}_2/\mathcal{W}_2)$. If $d_2 = \dim V_2$, $\mathcal{V}_2/\mathcal{W}_2 = 0$, so $h_1 = 0$. Otherwise we consider the restriction of h_1 to a slice $W_1 \times G_2$. On this variety, we have $h_1 \in H^0(G_2, Hom(\mathcal{W}_1, \mathcal{V}_2/\mathcal{W}_2))$, which is the same as $Hom(W_1, H^0(G_2, \mathcal{V}_2/\mathcal{W}_2)) = Hom(W_1, V_2)$, unless $d_2 = 0$, in which case we get 0. Now, letting W_1 vary, h_1 defines, and is determined by an element in $H^0(G_1, Hom(\mathcal{W}_1, \mathcal{V}_2)) = Hom(V_1, V_2)$. Now the same arguments applies to h_2, putting it in $Hom(V_2, V_1)$, unless $d_1 = dim V_1$, in which case $h_2 = 0$.

3.3 Theorem.

(a) **Let** $r < p$. *Let* $F_{r,p}$ *be the flag manifold*

$$F_{r,p} = \{(V, W); V \in M_{r,0}, W \in M_p, V \subset W\}. \tag{3.4}$$

Then we have the diagram

$$\begin{array}{ccc} F_{r,p} & \xrightarrow{\mu_r} & M_{r,0} \\ \nu_r \downarrow & & \\ M_p & & \end{array}$$

where μ, ν are the natural projections. This diagram describes M_r as the KN family of deformations of $Gr(r, C^p)$ in $M_{r,0}$.

(b) For $s < q$, we can similarly define $F_{s,q}$, and the corresponding projections μ, ν, describing M_s as the KN space for $Gr(s, C^q)$.

(c) Let $r < p, s < q$. Define $F_{r+s,p,q}$ as

$$F_{r+s,p,q} = \{(V, P, Q); V \in M_{r,s}, P \in M_p, Q \in M_q, V = (V \cap P) + (V \cap Q)\}.$$

Consider the diagram

$$\begin{array}{ccc} F_{r+s,p,q} & \xrightarrow{\mu_{r,s}} & M_{r,s} \\ \nu_{r,s} \downarrow & & \\ M_p \times M_q & & \end{array}$$

where $\mu_{r,s} = \mu_r + \mu_s$, and $\nu_{r,s} = \nu_r \times \nu_s$. This diagram describes $M_p \times M_q$ as KN space for $M_0 = M(C^p, C^q)$.

Proof. In each of these cases the proposition tells us that the candidate has the right dimension (by theorem 2.3), and thus is open in KN space. We shall show that it is closed in case (c); the other cases are just the same. Let $M(P_n, Q_n) \to M$. Then since the Grassmanians containing M_p and M_q are compact, we may assume that $P_n \to P$

in $\bar{M}_p$, and $Q_n \to Q$ in $\bar{M}_q$. Then $h|P$ is of signature $(p', 0)$ with $p' \leq p$, and $h|Q$ is of signature $(0, q')$ with $q' \leq q$. Now, by continuity, $M = M(P,Q)$ is contained in $M_{r,s}$, so the sum of every r-dimensional subspace of P with an s-dimensional subspace of Q is sof signature (r,s). This can only be if $p' = p, q' = q$, and thus $(P,Q) \in M_r \times M_s$.

We now turn to the Penrose correspondence in the case $rs \neq 0$. Let $d = r(p-r) + s(q-s)$. Let $T \to M_{r,s}$ be the tautological bundle of the Grassmanian $Gr(r+s, p+q)$ and E the associated determinant bundle. Let $E(-k)$ be the kth tensor power of E^{-1}. Let $\Theta(k) = R^d\nu_*\mu^*(E(-k))$. We recall the Penrose correspondence as defined by (2.14):

$$P : H^d(M_{r,s}, (E(-k)) \to H^0(M_p \times M_q, \Theta(k)).$$

We shall be comparing this to the case $rs = 0$, so we introduce this notation: a subscript '$+$' (as in $E_+(k)$) refers to bundles or sheaves on $M_{r,0}$ or M_p, and the subscript '$-$' refers to $M_{0,s}$ or M_q.

Now let $\nu : F \to M_p \times M_q$ be the KN family; F is the bundle over $M_p \times M_q$ whose fiber (over (P,Q)) is $\nu^{-1}(P,Q) = Gr(r,P) \times Gr(s,Q)$, and $\mu : F \to M_{r,s}$ is the map $\mu(L_+, L_-) = L_+ \oplus L_-$. As before, we let $M_0 = \nu^{-1}(C^p, C^q)$, and $M(C^p, C^q) = \mu(M_0)$. At the typical point W of M_0, with $W = L_+ \times L_-$, $\mu^{-1}(T)$ assigns the vector space $L_+ \oplus L_-$. Thus $\mu^{-1}(T) = T_+ \oplus T_-$ along M_0. Taking determinants, $\mu^{-1}(E(k)) = E_+(k) \times E_-(k)$. In order to understand the map P as defined by (2.14), we must, as in section 7 of [PR], consider the μ-vertical Dolbeault sequence of $\mu^{-1}(E(-k))$ on F. It suffices to calculate these exact sequences along $Gr(r, C^p) \times Gr(s, C^q)$; more precisely, these exact sequences are given by tensoring $E_+(k) \times E_-(k)$ with the μ-vertical exact sequence

$$0 \to \mu^{-1}(\mathcal{O}_{M_{r,s}}) \to \mathcal{O}_{K_0} \to \Omega^1_V \to \cdots \to \Omega^N_V \to 0 \tag{3.5}$$

We define the sheaves $Q_\pm$ by

$$0 \to T_\pm \to C^{p \ or \ q} \to Q_\pm \to 0 \tag{3.6}$$

and, for E a vector bundle E' denotes its dual.

3.7. Proposition. $\Omega^1_V = \Omega^1_+ \oplus \Omega^1_-$, *where* $\Omega^1_\pm$ *are defined by the exact sequences (the first map comes out of (3.6)):*

$$0 \to Q'_+ \otimes T_- \to (Q_+)^q \oplus (T_-)^p \to \Omega^1_+ \to 0$$

$$0 \to T_+ \oplus Q'_- \to (T_+)^q \oplus (Q'_-)^p \to \Omega^1_- \to 0 \tag{3.8}$$

.

Proof. Let W be a point in $M - 0 \in M_{r,s}, W = L_+ \oplus L_-$, with L_+ in C^p, L_- in C^q. We now describe $\mu^{-1}(W)$ in coordinates. F is the set of

$$(V, P, Q), V \in M_{r,s}, P \in M_p, Q \in M_q, \textit{ such that } V = (V \cap P) \oplus (V \cap Q).$$

$\{(A,B) \in Hom(C^p, C^q) \oplus Hom(C^q, C^p)\}$ coordinatizes a neighborhood of

$$P_0 \in \nu(\mu^{-1}(W))$$

as $\{(P_A, Q_B)\}$ where

$$P_A = \{(z, Az)\}, \quad P_B = \{(Aw, w)\}.$$

(V, P_A, P_B) is in $\mu^{-1}(W)$ if and only if $P_A \cap W$ has dimension r, and $P_B \cap W$ has dimension s. If (z, Az) is in $P_A \cap W$, we must have $z \in L_+$ and $Az \in L_-$. For this space to have dimension r, we must have $A : L_+ \to L_-$. Similarly for B, so

(3.9). $$\mu^{-1}(W) = \{(A,B); A : L_+ \to L_-, B : L_- \to L_+\}$$

Since this is a linear space it parametrizes the space of vertical tangents, and thus describes the dual Ω^1_V. Thus we have the exact sequence defined by (3.9):

$$0 \to \Omega^1 \to Hom(C^p, C^q) \oplus Hom(C^q, C^p) \to Hom(T_+, C^q/T_-) \oplus Hom(T_-, C^p/T_+) \to 0.$$

This really is two exact sequences, giving the splitting $\Omega^1 = \Omega^1_+ \oplus \Omega^1_-$, corresponding to the kernels of the maps

$$Hom(C^p, C^q) \to Hom(T_+, C^q/T_-) \to 0$$

for Ω^1_+, and similarly for Ω^1_-. Otherwise put we have these sequences defining $\Omega^1_\pm$:

(3.10) $$0 \to \Omega^1_+ \to C^{pq} \to T' - + \otimes Q_- \to 0,$$

(3.11) $$0 \to \Omega^1_- \to C^{pq} \to Q_+ \otimes T'_- \to 0,$$

Notice that if we tensor together the two sequences

$$0 \to Q'_+ \to C^p \to T'_+ \to 0, \quad 0 \to T_- \to C^q \to Q_- \to 0,$$

we get

$$0 \to Q'_+ \otimes T_- \to (Q'_+ \otimes C^q) \oplus (T_- \otimes C^p) \to C^{pq} \to T'_+ \otimes Q_- \to 0.$$

Comparing this with (3.10) gives us the exact sequence (3.8).

3.12. Theorem. *If $k \geq min(p,q)$, the Penrose correspondence is as follws:*

(3.13) $$P : H^d(M_{r,s}, E(-k)) \to H^0(M_p, \Theta(k)) \otimes H^0(M_q, \Theta(k)).$$

If $k \geq max(p,q)$, this map is injective.

Proof. (a) As we have seen $\mu^*(E(-k)) = E_+(-k) \otimes E_-(-k)$, so P maps into $R^d\nu_*(E_+(-k) \otimes E_-(-k))$: the space of sections of the vector bundle (over $M_p \times M_q$) whose fiber is $H^d(M(P,Q), E_+(-k) \otimes E_-(-k))$. But this is the same as

$$H^{d_1}(Gr(r,P), E_+(-k)) \otimes H^{d_2}(Gr(s,Q), E_-(-k)),$$

since all other terms of the Kunneth formula will vanish. (b) From the preceding proposition, we have that Ω_V^k is given by

$$\oplus_{u+v=k}\Lambda^u(\Omega_+)\otimes\Lambda^v(\Omega_-)$$

Using the identifications of proposition 3.7, the injectivity is proven by a calculation as that in theorem 7.14 of [8].

Now, to understand P as a mapping of $SU(p,q)$-modules, we have to turn to the nonholomorphic family

$$\begin{array}{ccc} & G/H\cap K & \\ \swarrow\mu & & \searrow\nu \\ M_{r,s} & & M_p \end{array}$$

and define the restricted Penrose transform

3.15 $$P_0 : H^d(M_{r,s}, E(-k)) \to \Gamma^\omega(M_p, R^d\nu_*\mu^*(E(-k)),$$

where by Γ^ω we mean the space of real analytic sections. This P_0 is an intertwining operator. Now the map $\alpha : M_p \to M_p \times M_q$ given by $P \to (P, P^\perp)$ is the map which induces the family above, so that $P_0 = \alpha^*(P)$; P_0 is the restriction of P to $\{(P,Q); Q = P^\perp\}$. Thus, we conclude:

3.16. Theorem. *$P_0 : H^d(M_{r,s}, E(-k)) \to H^0(M_p, \Theta(k)) \otimes H^0(M_q, \Theta(k))$ is an intertwining operator of the natural action of $SU(p,q)$ on $M_{r,s}$ with the natural action on the first factor and the adjoint action in the second factor. Thus, these representations of $SU(p,q)$ lie in the tensor product of a holomorphic and an antiholomorphic representation.*

Bibliography

[1] A. . Andreotti and F. Norguet, Problème de Lewy et convexité holomorphe pour les classes de cohomologie, *Ann. Scuola Norm. Sup. Pisa* **20** (1966), 197-241.

[2] R. J. Blattner and J. H. Rawnsley, Quantization of $U(k,l)$ on $R^2(k+l)$,*J. Funct. Anal.* **50** (1983), 188-214.

[3] M. G. Eastwood, The generalized twistor transform and unitary representations of $SU(p,q)$, *preprint.*

[4] M. G. Eastwood, R. Penrose and R. O. Wells, Cohomology and massless fields, *Comm. Math. Physics* **78** (1981), 305-351.

[5] K. Kodaira, A theorem of completeness of characteristic systems for analytic families of compact submanifolds of complex manifolds, *Ann. of Math.* **75** (1962), 146-162.

[6] L. A. Mantini, An analog of the Penrose correspondence for representations of $U(p,q)$ on L^2-cohomology, *Thesis*, Harvard University, 1983.

[7] M. Namba, On maximal families of compact submanifolds of complex manifolds, *Tohôku Math. Journal* **24** (1972), 581-609.

[8] C. Patton and H. Rossi, Unitary structures on cohomology, *to appear in T. A. M. S. (1985)*.
[9] J. H. Rawnsley, W. Schmid and J. A. Wolf, Singular unitary representations and indefinite harmonic theory, *J. Funct. Anal.* (1983).
[10] W. Schmid, Homogeneous complex manifolds and representations of semisimple Lie groups, *Thesis*, Univ. of Cal., Berkeley, 1967.
[11] R. O. Wells and J. Wolf, Poincaré series and automorphic cohomology on flag domains, *Ann. of Math.* **105** (1977), 397-448.

Hewlett-Packard Corvallis, OR 97331 , U.S.A.

Mathematics Department University of Utah Salt Lake City, UT 84103 , U.S.A.

Contemporary Mathematics
Volume 58, Part I, 1986

SOME APPLICATIONS OF THE LEFSCHETZ FIXED POINT THEOREMS IN (COMPLEX) ALGEBRAIC GEOMETRY

C.A.M. Peters

ABSTRACT. Lefschetz fixed point theorems are applied to determine the singularities of the moduli space of curves (this simplifies existing methods) and to the fixed point set of involutions on K3-surfaces and Enriques surfaces.

INTRODUCTION.

There are various versions of Lefschetz' fixed point theorem in complex-analytic geometry and algebraic geometry. Applications are not new, the most striking being of course its use as a strategy to prove the Weil conjectures. I shall restrict myself to the complex-analytic category however.

After a short résumé of what I need from the various Lefschetz fixed point theorems in §1, I give an application to the study of the singularities of the moduli space of curves. This yields a short proof of a result originally obtained by Rauch in [R]. In §3 I apply the holomorphic Lefschetz fixed point formula to involutions of K3-surfaces and Enriques surfaces.

For an application in the same spirit I refer to [B-P-V, Theorem VIII, (21.4)].

For this paper I have profitted very much from stimulating conversations with participants of the Conference.

§1. THE FIXED POINT THEOREMS AND TWO APPLICATIONS.

For any continuous self-map f of a compact oriented manifold X, the intersection number of the graph of f and the diagonal in $X \times X$ is equal to the topological Lefschetz-number

$$L_{top}(f) = \sum_q (-1)^q \operatorname{Trace}(f^* : H^q(X) \to H^q(X)),$$

1980 Mathematics Subject Classification. 58G 10, 14D22, 14H15, 14J25

where $H^q(X)$ denotes the q-th cohomology group of X with coefficients in a field.

If X in addition is a complex manifold and f a holomorphic map the aforementioned intersection number can be computed in an important special case, namely when the connected components Y of the fixed locus of f are non-degenerate, i.e. Y is a manifold and at every point $P \in Y$ the linear map $(1 - df_p)$ is invertible, when restricted to the co-normal space at $P \in Y$.

COROLLARY. If X is a compact complex manifold and f a holomorphic self-map of X with non-degenerate fixed locus, we have

$$L_{top}(f) = \sum_Y \text{euler number}(Y)$$

This holds in particular in case $f^n = id_X$ for some $n \geq 1$.

For a proof we refer to [Ue]. The fact that any f with finite order has non-degenerate fixed point locus follows e.g. from [C].

In the complex-analytic category more refined theorems are valid such as the holomorphic Lefschetz fixed point formulas (cf. e.g. [A-B], [A-S], [A-Se], [T-T]). We shall use only two very special cases of the following general situation.

We have a compact complex manifold X, a holomorphic vectorbundle V on X, a holomorphic self map f of X and a holomorphic bundle map $\phi : f^*V \to V$. Moreover we assume that the fixed locus of f is non-degenerate. We set

$$L(f,\phi) := \sum_p (-1)^p \text{ Trace } f^* \circ \phi : H^p(X,V) \to H^p(X,f^*V).$$

In the special case where $V = \mathcal{O}_X$, the trivial line bundle on X and $\phi = f^* : \mathcal{O}_X \to \mathcal{O}_X$, we set

$$L(f,\phi) = L_{hol}(f)$$

In this general setting it is possible to associate complex numbers ν_Y to each component Y of the fixed locus of f such that $L(f,\phi) = \sum_Y \nu_Y$, see [T-T] for details. An explicit computation for the ν_Y's can be found in [A-S]. We only need the following

FACTS

1) If $\sum_Y \nu_Y \neq 0$, f has at least one fixed point,

2) If all Y's are points P we have

$$L(f,\phi) = \sum_P \frac{\mathrm{Trace}\ \phi_P}{\det(1-(f_*)_P)}$$

([A-Se,p.543],[A-B,p.459])

3) If $\dim_{\mathbb{C}} X = 2$ and f is an involution we have

$$L_{hol}(f) = \tfrac{1}{4}(\text{number of fixed points}) - \tfrac{1}{4} \sum_D (K_X \cdot D),$$

where D runs over the 1-dimensional components of the fixed locus of f. ([A-S, p.568])

Let me close this section with a few applications to automorphisms inducing the identity in cohomology.

APPLICATION 1. Let f be a holomorphic automorphism of a compact Kähler manifold X inducing the identity in cohomology. If X has no holomorphic fields then its fixed point set has the same Euler number as X.

Indeed, since X is Kähler it follows e.g. from [L] that f has finite order, hence its fixed locus is non-degenerate, so that the above Corollary applies. □

For a more detailed investigation in case $\dim X \leq 2$ we refer to [P1] and [P2].

APPLICATION 2. (Atiyah-Bott, [A-B, II, Cor. 4.13]). If $H^{0,q}(X) = 0\ \forall q > 0$, every holomorphic self map of (the complex manifold) X has a fixed point.

This follows from Fact 1 applied to $L_{hol}(f) =$

$$= \sum_p (-1)^p\ \mathrm{Trace}\ (f^* | H^p(\mathcal{O}_X) \to H^p(\mathcal{O}_X)) = 1\ .$$

§2. AN APPLICATION TO THE MODULI VARIETY OF CURVES OF GENUS $g \geq 2$.

A well known procedure to determine the singularities of the coarse moduli M_g of curves of genus g is to investigate the action of automorphisms of a curve C on the tangent space of the Kuranishi space of G, identified with $H^1(\Theta_C)$. Indeed, if $c \in M_g$ denotes the point corresponding to C, a neighbourhood of c in M_g is isomorphic to the quotient by $G = \mathrm{Aut}(C)$ of a neighbourhood of the origin in $H^1(\Theta_C)$. Since $g \geq 2$, G is a finite group we may detect whether c is singular by applying the following

CRITERION. Suppose a group G acts properly discontinuously on a complex manifold M and suppose that the fixed point set $\mathrm{Fix}(G) = \{m \in M;$ stabilizer in G of $m \neq$ identity$\}$ has codimension ≥ 2 in M. Then $m \in M$ maps to a singularity of M/G if and only if $m \in \mathrm{Fix}(G)$.

For a proof see e.g. [P-S, Lemma (9.11]. Observe that the condition implies that G should act effectively, so in applying the criterion, we should replace G by $G/$(hyperelliptic involution) in case $g=2$ (so we exclude the hyperelliptic involution in this case). Let us see when the condition in the above criterion is satisfied.

THEOREM. Let C be a Riemann-surface of genus $g \geq 2$ and suppose that $\gamma \in \mathrm{Aut}\, C$ fixes a 1-codimensional subspace of $H^1(\Theta_C)$, then either $g=3$, C is hyperelliptic and some power of γ is the hyperelliptic involution or $g=2$, and some power of γ is an involution such that $C/\{1,\gamma\}$ is elliptic.

PROOF. We apply Fact 2 of §1 to the tangent bundle Θ_C. We may assume, replacing γ by some power, that $\mathrm{ord}(\gamma)$ is a prime number ℓ and we find

$$\sum_{k=1}^{\ell-1} L(\gamma^k, \gamma_*^k) = (\#\text{fixed points}) \sum_{k=1}^{\ell-1} \frac{\lambda^k}{1-\lambda^k} \quad , \lambda = e^{\frac{2\pi i}{\ell}}$$

since $(\gamma_*^k)_p$ is multiplication by some ℓ-th root of unity, and for $k=1,\ldots,\ell-1$ all of them occur.

The left hand side equals

$$\ell \,\Sigma\, (-1)^i \dim H^i(\Theta_C)^{\mathrm{invar}} - \chi(\Theta_C) = -\ell \dim H^1(\Theta_C)^{\mathrm{inv}} + \dim H^1(\Theta_C).$$

Assuming that γ fixes a hyperplane, we get

$$-\ell(\dim H^1(\Theta_C) - 1) + \dim H^1(\Theta_C) = (1-\ell)(3g-3) + \ell .$$

The right hand side equals $-\frac{1}{2}(\ell - 1)$ ($\#$fixed points), hence we find for the number of fixed points $6(g-1) + \frac{2\ell}{\ell-1}$. Since this must be a non-negative integer we find $\ell=2$ or 3. In case $\ell=2$ Hurwitz formula easily gives that either $g=3$ and $X/\gamma \cong \mathbb{P}_1$ or $g=2$ and X/γ is elliptic, whereas $\ell=3$ is seen to contradict Hurwitz formula. □

COROLLARY. Suppose we are in characteristic zero. If $g \geq 4$, M_g is singular at c if and only if the corresponding curve has non-trivial automorphisms. For $g=3$ this criterion holds on $M_3 \setminus$ (hyperelliptic locus) and for $g=2$ outside the divisor of hyperelliptic-elliptic curves. (If $g=2$ "non-trivial" means "different from the identity or the hyperelliptic involution).

REMARKS. 1) For $g=2$ a much more precise theorem is known valid in all characteristics, see [I]. It follows from his results that M_2 has a unique singular point corresponding to the hyperelliptic-elliptic curve with equation $y^2 = x^6 - x$.

2) For $g=3$, the same proof gives that the hyperelliptic locus is singular at points whose corresponding curves have more than 2 automorphisms (the trivial one and the hyperelliptic involution). But in fact it is known that M_3 itself is singular at these points (see [Rauch]).

3) The results for $g \geq 4$ have been shown by [R] in characteristic zero. Our proof is considerably shorter.
For positive characteristics one should consult [Po], [O].

§3. APPLICATIONS TO INVOLUTIONS OF K3-SURFACES AND ENRIQUES SURFACES.

A K3-surface is a compact complex surface which is simply connected and whose canonical bundle is trivial. If a K3-surface admits a fixed point free involution the quotient by definition is an Enriques surface. For details we refer to [B-P-V, Ch. VIII], we only use the following
FACTS:

1) The Betti-numbers of a K3-surface are $b_0 = b_4 = 1$, $b_1 = b_3 = 0$, $b_2 = 22$; for an Enriques surface one has $b_0 = b_4 = 1$, $b_1 = b_3 = 0$, $b_2 = 10$.

2) The canonical class is numerically trivial on both kinds of surfaces and therefore an application of the adjunctionformula gives:

3) If D is a smooth curve on a K3-surface or an Enriques surface its genus π_D equals $\frac{1}{2}D^2 + 1$.

4) An Enriques surface has no non-zero holomorphic 1-forms or 2-forms.

From 2) and §1 we find for an involution ι of a K3-surface or Enriques surface X:

$$(1) \begin{cases} L_{top}(\iota) = \nu(\iota) - \sum_D D^2 = \nu(\iota) - 2\sum_D(\pi_D - 1) \\ L_{hol}(\iota) = \frac{1}{4}\nu(\iota), \end{cases}$$

where $\nu(\iota)$ = number of fixed points of ι and D runs over the fixed curves of ι.

From 4) we find $L_{hol}(\iota) = 1$, hence (1) gives:

PROPOSITION 1. An involution on an Enriques surface has 4 isolated fixed points (and possibly some fixed curves).

If we let b_2^{inv}, resp. b_2^{anti} be the rank of the invariant, resp. anti-invariant sublattice of $H^2(X, \mathbb{Z})$, we have:

$$(2) \quad L_{top}(\iota) = 2 + (b_2^{inv} - b_2^{anti}) = 2 - b_2 + 2b_2^{inv}$$

and Fact 1) gives

PROPOSITION 2. If ι is an involution on an Enriques surface X and D are the fixed curves of ι we have

$$\Sigma_D(\pi_D - 1) = 6 - b_2^{inv} .$$

COROLLARY. If $\pi_D \leq 1$ for all D, there can be at most 4 rational D; if - on the contrary there exists a fixed curve with genus ≥ 2 this curve is unique. If π is its genus we have $\pi \leq 6$.

PROOF. If $\pi_D \leq 1$ for all D, $6 - b_2^{inv} \geq -4$, hence the formula of Proposition 2 shows that there can be at most 4 rational D.

Now, remark that N disjoint curves give N independent cohomology classes provided all self intersections are non-zero. Secondly, at most one among them can have positive self-intersection, because of the index-theorem [B-P-V,p.122]. By the last remark there can be at most one D with $\pi_D \geq 2$ and if there are k further rational fixed curves, the first remark shows that $b_2^{inv} \geq k+1$, hence the formula of Proposition 2 shows that $-k + \pi - 1 \leq 5-k$, hence $\pi \leq 6$. □

EXAMPLES

1) If ι has 4 rational fixed curves and no other fixed curve of genus ≥ 2 one must have $b_2^{inv} = 10$, i.e. ι acts trivially in cohomology. Surfaces with this property have been classified by Mukai and Namikawa [M-N]. They show that there are two types of examples. Upon inspecting them one finds:
   - In the first example the four rational curves on S that are images of the curves P_{x_1}, Q_{x_2} are fixed under σ (notation: see §4 of [M-N]) and no other curves.

- In the second example the four rational curves on S that are the images of the curves $\tilde{X}_{\pm}, \tilde{Y}_{\pm}, \tilde{F}_{\pm,\pm}$ and the elliptic curve that is the image of $\tilde{E}$ are fixed under σ and no other curves (notation from §4 of [M-N]).

2) It is easy to give examples for each π with $2 \leq \pi \leq 5$, namely: Every Enriques surface can be represented as a double cover of a 4-nodal quartic Del Pezzo surface Q branched along $\gamma \cup \operatorname{Sing} Q$, where $\gamma \in |\mathcal{O}_Q(2)|, \gamma \cap \operatorname{Sing} Q = \emptyset$. Such a γ has arithmetic genus 5, hence the imposition of k double points $0 \leq k \leq 3$ gives a branch curve of genus 5-k. (See [Co,§7]).
If one takes for γ a reducible curve one can produce 1,2,3 or 4 rational fixed curves (the case with 4 curves being identical with example 1).

3) I don't know whether π can be 6.

We next consider K3 surfaces. If X is a K3-surface and ω_X is any nowhere zero holomorphic 2-form (unique up to multiplication by a non-zero complex number) an involution ι of X either preserves ω_X or maps it to $-\omega_X$.

CASE a) $\iota^*(\omega_X) = \omega_X$.

In the case one sees either directly or from the holomorphic Lefschetz fixed point formula (§1 Fact 3) that all fixed points of ι are isolated and that there are 8 of them. The quotient $X/\{1,\iota\}$ is again a K3-surface. Such involutions are known as "Nikulin-involutions". See [N, section 5], [M, section 5].

CASE b) $\iota^*(\omega_X) = -\omega_X$.

Now all fixed points are distributed over <u>curves</u> and one has

PROPOSITION 3. An involution ι on a K3-surface which does not leave invariant the holomorphic 2-forms has at most fixed curves D and $\sum_D(\pi_D - 1) = 10 - b_2^{inv}$.

PROOF. Immediate from (1),(2) and Fact (2).

COROLLARY. If $\pi_D \leq 1$ for all D, there can be at most 10 rational curves. If some fixed curve of genus ≥ 2 occurs among the D, this curve is unique and has genus ≤ 10.

PROOF: similar to the proof of the Corollary to Prop. 2.

□

EXAMPLES. Let $\overline{X}$ be a double covering of $\mathbb{P}_2$ branched in a sextic having at most simple singularities. The minimal resolution X of $\overline{X}$ is a K3-surface [B-P-V, p. 184] and the covering involution fixes the proper transform of the sextic plus certain resolution curves. If the sextic is irreducible with k ordinary double points, its genus is 10-k and so for each possible value of π_D we can construct examples. If we use reducible sextics and more complicated singularities we can construct examples with k fixed rational curves for each value of k with $0 \le k \le 10$. With k lines in general position we can realize every k up to 6, for $k=7,8,9,10$ we take 6 lines such that there are exactly $k-6$ points through which precisely 3 lines pass. The situations $k=9,10$ are sketched below

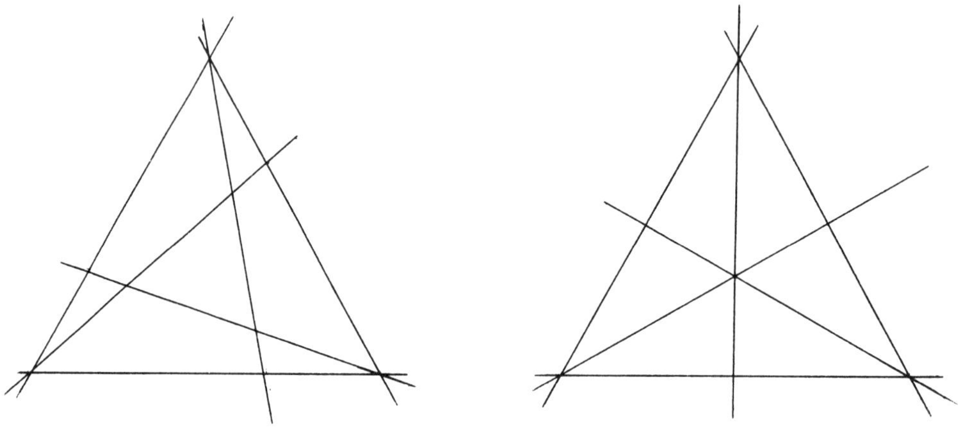

(This has been pointed out to me by F. Hirzebruch).
It is an interesting problem to classify all possible involutions according to their fixed point locus.

BIBLIOGRAPHY.

[A-B] M.F. Atiyah, R. Bott, A Lefschetz fixed point formula for elliptic complexes I, Ann. Math. 86 (1967), 347-407, ibid, II, Ann. Math. 88 (1968), 451-491.

[A-S] M.F. Atiyah, I. Singer, The index of elliptic operators III, Ann. Math. 87 (1968), 546-604.

[A-Se] M.F. Atiyah, G. Segal, The index of elliptic operators II, Ann. Math. 87 (1968), 531-545.

[B-P-V] W. Barth, C. Peters, A. Van de Ven, Compact complex surfaces, Ergebnisse der Math. 3.Folge 4, Springer-Verlag 1984, Berlin etc.

[C] H. Cartan, Quotient d'un espace analytique par un groupe d'automorphismes, in: Algebraic Geometry, a symposium in Honor of S. Lefschetz, 90-102. Princeton University Press, Princeton 1957.

[Co] F. Cossec, Projective models of Enriques surface, Math. Ann. 265(1983), 283-334.

[I] J.I.-Igusa, Arithmetic variety of moduli for genus two, Ann. Math. 72 (1960), 612-649.

[L] D. Lieberman, Compactness of the Chow scheme: applications to automorphisms and deformations of Kähler manifolds, Sém. Norguet 1976, Lect. Notes in Math. 670, Springer Verlag 1978.

[M] D.R. Morrison, On K3 surfaces with large Picard number, Inv. Math. 75 (1984), 105-121.

[M-N] S. Mukai, Y. Namikawa, Automorphisms of Enriques surfaces which act trivially on the cohomology groups, Inv. Math. 77(1984), 383-398.

[N] V. Nikulin, Finite groups of automorphisms of Kählerian surfaces of type K3, Trudy Mosk. Mat. 06.38, 75-137 (1979), Trans. Moscow Math. Soc. 38, 71-135 (1980).

[O] F. Oort, Singularities of the moduli scheme for curves of genus three, Proc. Kon. Ned. Akad. Wetenschappen 78 (1975) (=Indag. Math. 37 (1975)), 170-174.

[P1] C.A.M. Peters, Holomorphic automorphisms of compact Kähler surfaces and induced actions in cohomology, Inv. Math. 52 (1979), 143-148.

[P2] C.A.M. Peters, On automorphisms of compact Kähler surfaces, in: Journées de gêomêtrie algébrique d'Angers (juillet 1979), Sijthoff & Noordhoff 1980, Alphen aan den Rijn, Netherlands, 249-267.

[Po] H. Popp, The singularities of the moduli scheme of curves, Journ. number theory 1 (1969), 90-107.

[P-S] Chr. Peters, J. Steenbrink, Infinitesimal variations of Hodge structure and the generic Torelli problem for projective hypersurfaces (after Carlson, Donagi, Green, Griffiths, Harris), in: Classification of algebraic and analytic manifolds, Progress in Math. 39, Birkhäuser Verlag (1983).

[R] H.E. Rauch, The singularities of the modulus spaces, Bull. Am. Math. Soc. 68(1962), 390-394.

[T-T] D. Toledo, Y.L.L. Tong, The holomorphic Lefschetz formula, Bull. Am. Math. Soc. 81 (1975), 1133-1135.

[Ue] K. Ueno, A remark on automorphisms of Enriques surfaces, Journ. Fac. Sc. Univ. Tokyo, Sec. IA, 23 (1976), 149-165.

<u>Remark</u> (May 21, 1985)
In regard to the point 3 on page 7, recently Ulf Persson and I have proved that $\pi \neq 6$

DEPARTMENT OF MATHEMATICS
LEIDEN UNIVERSITY
LEIDEN, WASSENAARSEWEG 80, NETHERLANDS.

Contemporary Mathematics
Volume 58, Part I, 1986

STABILITY OF TWO-DIMENSIONAL LOCAL RINGS. II

Jayant Shah[1]

DEFINITION. If R is a local ring of dimension r and m is a positive integer, then $e_m(R)$, the mth flat multiplicity of R is defined by

$$e_0(R) = \sup \left\{ \frac{e(I)}{r!\operatorname{col}(I)} : \ I \text{ of finite colength in } R \right\}$$

$$e_m(R) = e_0(R[[t_1,\ldots,t_m]])$$

in which e(I) denotes the multiplicity of the ideal I.

DEFINITION. A local ring R is called semistable if $e_1(R) = 1$; R is called stable if, an addition, the defining sup is not attained.

These definitions are due to Mumford. We refer to [1] and [2] for the motivation behind these definitions and earlier results. In particular, stability of surface singularities for which multiplicity = embedding dimension and those for which multiplicity = embedding dimension - 1 $\leqslant 3$ are fairly well understood. Here, we announce some preliminary results concerning surface singularities for which multiplicity = embedding dimension - 1 = 4.

Let R be a Cohen-Macaulay local ring of dimension 2 over an algebraically closed field k.

THEOREM. Suppose that R is semistable and that its multiplicity = 4 and the embedding dimension = 5. Then the projective tangent cone of R is isomorphic to either the projective tangent cone of a quotient singularity [2] or one of the following one-dimensional schemes in $\mathbb{P}_4$:

(a) 4 lines meeting at a quadruple point of embedding dimension 4,

(b) a planar double line and two other lines meeting at a quadruple point of embedding dimension 4,

(c) a double conic spanning $\mathbb{P}_4$,

(d) two planar double lines meeting transversally at a quadruple point of

[1]Partially supported by a grant from the National Science Foundation.

0271-4132/86 $1.00 + $.25 per page

embedding dimension 4.

The method of proving the theorem is the same as in [3]. We first list all possible arithmetically Cohen-Macaulay, one dimensional schemes of degree 4 in $\mathbb{P}_4$ and write down their defining equations. These give us the initial forms of the defining equations of R. For really bad singularities, these equations are enough to estimate $e_1(R)$ and show that R must be unstable. We then write down syzygies to analyze the remaining cases.

BIBLIOGRAPHY

[1] Mumford, D., Stability of projective varieties, L'Enseignement Mathématique, T. XXIII, fasc. 1-2, 39-110 (1977).

[2] Riemenschneider, O., Zweidimensionale Quotientensingularitäten: Gleichungen und Syzygien, Arch. Math. 37, 406-417 (1981).

[3] Shah, J., Stability of two-dimensional local rings. I. Invent. math. 64, 297-343 (1981).

DEPARTMENT OF MATHEMATICS
NORTHWESTERN UNIVERSITY
BOSTON, MASS. 02115
U.S.A.

Contemporary Mathematics
Volume 58, Part I, 1986

THE CONNECTION BETWEEN LINEAR SERIES ON CURVES AND GAUSS MAPS ON SUBVARIETIES OF THEIR JACOBIANS

Roy Smith(*)
Horacio Tapia-Recillas(**)

It is classical that linear series on curves are linked to singularities on special subvarieties of Jacobians by the celebrated Riemann-Kempf "singularity theorem". In this paper we show there is also a close connection between the linear series on a curve and the geometry of the smooth points of these subvarieties, provided by the Gauss map. There is in fact more precise geometric information encoded in the Gauss maps than in the tangent cones to the singularities, the tangent cones being (in the simplest cases) just the images of (certain restrictions of) the Gauss maps. Also the Gauss maps reflect more closely the rational maps associated to linear series and thus detect the presence of base loci. Furthermore, whereas the singularities theorem applies only to linear series of degree $\leq g-1$, the Gauss map technique applies to all special linear series. The present methods allow one to obtain "refined Torelli theorems" as well, i.e. to reconstruct both the curve and the special linear series on the curve directly from the geometry of the theta-divisor.

The discussion here is an extension of the proof of Torelli's theorem given by Andreotti, which from our point of view is a reconstruction of the pair (C,ϕ_K) consisting of the curve plus the rational map associated to its canonical series, from the Gauss map on the theta-divisor.

The main result in this paper deals with the recovery of a curve and its special g^1_d's from the Gauss map on W_{d-1}, (notations are explained in section §0 and a more complete statement is in section §2):

THEOREM. If C has a g^1_d but no g^1_{d-1} and no g^2_{d+1}, for some d with $d \leq g = \text{genus}(C)$, if $\phi: W_{d-1} \to \mathbb{G}r(d-2,g-1)$ is the canonical map on W_{d-1}, if

(*) partially supported by NSF grant #DMS-8317078
(**) partially supported by CONACyT grant #PCCBND-001373

R is the multiple locus of ϕ, and if $B = \phi(R)$, then the restriction $\phi : R \to B$ is a "universal" $\phi_{g^1_d}$ for C.

This work arose from our collaboration at the Centro de Investigación y de Estudios Avanzados del I.P.N. in Mexico in 1980 and was outlined in a talk by the second named author in the Harvard algebraic geometry seminar in 1981.

We are grateful to the organizers of the Lefschetz Centennial Conference for the opportunity to present it here in more detail. The "refined Torelli problem" posed to us by Phillip Griffiths was an important stimulus to this work, and we are grateful both to him and to David Mumford for enlightening conversations on these ideas. The present article is devoted primarily to the study of g^r_d's with $r = 1$. The general case will be the subject of a second article.

§0. PRELIMINARIES

We begin by recalling the general setting, primarily the structure of the Abel map and its derivative, the Gauss map. Andreotti's paper [1] is the reference for this section.

Assume C is a smooth connected curve of genus g over the complex numbers, let $C^{(d)}$ denote the d-fold symmetric product of the curve, let $J(C)$ be the Jacobian variety of C, and let $\alpha : C^{(d)} \to J(C)$ be the Abel map

$$\alpha(P_1,\ldots,P_d) = < \sum_{i=1}^{d} \int_{P_0}^{P_i} \omega_1, \ldots, \sum_{i=1}^{d} \int_{P_0}^{P_i} \omega_g >,$$

defined by a basis $\{\omega_1,\ldots,\omega_g\}$ of the global holomorphic one-forms on C, and by a base point $P_0 \in C$. The image $W_d = \alpha(C^{(d)})$ is an algebraic subvariety of dimension d in $J(C)$, for $1 \leq d \leq g$. As illustrated by the following diagram,

$$\begin{array}{ccc} C & \xrightarrow{\alpha_*} & \mathbb{P}^{g-1} \cong \mathbb{P}(T_0(J(C))) \\ {\scriptstyle\alpha}\searrow & & \nearrow{\scriptstyle\gamma} \\ & W_1 & \end{array}$$

the Gauss map γ on W_1, which assings to $x \in W_1$, the tanget space $T_x(W_1)$, is equivalent to the projectivized derivative α_* of the Abel map, which in turn is just the canonical map ϕ_K on the curve.

That is, if $\alpha(p) = \langle \int_{p_0}^{p} \omega_1, \ldots, \int_{p_0}^{p} \omega_g \rangle$ then the derivative $\alpha_{*,p}$ is a linear map $\alpha_{*,p}: T_p(C) \to T_{\alpha(p)}(J(C)) \cong T_0(J(C))$ and the projectivized derivative α_* is given by taking its image,

$$\begin{aligned} \alpha_*(p) = \alpha_{*,p}(T_p(C)) &\in \mathbb{G}r(1, T_0(J(C))) \\ &\cong \mathbb{G}r(0, \mathbb{P}(T_0(J(C)))) \\ &\cong \mathbb{P}(T_0(J(C))) \\ &\cong \mathbb{P}^{g-1} \end{aligned}$$

Then since the canonical map is given by

$$\begin{aligned} \phi_K(p) = \langle \omega_1(p), \ldots, \omega_g(p) \rangle &\in \mathbb{P}(H^0(C;K)^*) \\ &\cong \mathbb{P}(T_0(J(C))), \end{aligned}$$

the fundamental theorem of calculus implies that $\phi_K(p) = \alpha_*(p)$. More generally, for $1 \leq d \leq g-1$ the analogous diagram commutes,

$$\begin{array}{ccc} C^{(d)} & \overset{\alpha_*}{\dashrightarrow} & \mathbb{G}r(d-1, \mathbb{P}(T_0(J(C)))) \\ & {\scriptstyle\alpha}\searrow \quad \nearrow{\scriptstyle\gamma} & \\ & W_d & \end{array}$$

and α_* is again equal to the canonical map ϕ_K on $C^{(d)}$ [1,p.811]. In this case α_* and γ are only rational maps, γ is defined on the locus of smooth points of W_d, $W_d\text{-sing}\cdot(W_d)$, and α_* is defined on the set $\alpha^{-1}(W_d\text{-sing}\cdot(W_d))$.

More precisely, if $D = (p_1, \ldots, p_d) \in C^{(d)}$ is a point such that $h^0(C;D) = 1$, then ϕ_K is defined at D, $\alpha(D)$ is a non-singular point of W_d, the images of the points $\{P_1, \ldots, P_d\}$ on the canonical curve

$$\phi_K(C) = \Gamma \subseteq \mathbb{P}(T_0(J(C)))$$

are in general position ("geometric Riemann-Roch" [3,p.248], and $\phi_K(D)$ = the $(d-1)$-dimensional subspace of $\mathbb{P}(T_0(J(C)))$ represented by $T_{\alpha(D)}(W_d)$, which is exactly the one spanned by the d points $\{\phi(P_1), \ldots, \phi(P_d)\}$. We remark that the "span" of these points is defined as the base locus of the set of hyperplanes $H \subseteq \mathbf{P}^{g-1}$ such that as divisors on C, $\phi_K^{-1}(H) = H\cdot C \geq \sum_1^d P_i$, and of course we say the points $\{P_1,\ldots,P_d\}$ are in general position if this base locus has dimension $d-1$. Thus with these conventions, we may write

$$T_{\alpha(D)}(W_d) = \phi_K(P_1,\ldots,P_d) = \overline{\{\phi_K(P_1),\ldots,\phi_K(P_d)\}}$$

[1,pp.810-812], where a bar written over a set (or a divisor) denotes its span.

§1. THE FUNDAMENTAL EXAMPLE

We will now describe in detail the simplest example, the Gauss map on W_2 for a curve of genus 4, which we assume at first to be non-hyperelliptic. Then C is the complete intersection of a quadric and a cubic surface in $\mathbf{P}^3$ and has either one or two g_3^1's according to whether the quadric has rank 3 or 4, [4,p.55]. We will recover C and the g_3^1's from the Gauss map on W_2.

Consider the diagram:

$$\begin{array}{ccc} C^{(2)} & \overset{\alpha_*}{\dashrightarrow} & \mathbb{G}r(1,3) \\ & {\scriptstyle\alpha}\searrow \quad \nearrow{\scriptstyle\gamma} & \\ & W_2 & \end{array}$$

in which, if $(p,q) \in C^{(2)}$ then $\gamma(\alpha(p,q)) = \alpha_*(p,q)) = \overline{pq}$ = the secant line determined by p and q on the canonical model of C in $\mathbf{P}^3$.

PROPOSITION. γ is a finite map, birational onto its image, having a multiple locus in W_2 consisting of either one or two components, each birational to the curve C, and on which γ restricts to the rational maps

$$\phi_L : C \to \mathbf{P}^1$$

defined by the one or two linear series $L = g_3^1$ on C.

PROOF. C has no g_2^1 by hypothesis, so α is injective by the converse of Abel's theorem, and W_2 is non-singular by the Riemann-Kempf singularity theorem. Consequently, by Zariski's main theorem α is an isomorphism and we may use it to identify $C^{(2)}$ with W_2, and γ with α_*.

Then for $(p,q) \in C^{(2)}$, $\gamma^{-1}(\gamma(p,q))$ = all pairs of points (r,s) on C such that the secants $\overline{rs}$ and $\overline{pq}$ are equal. By geometric Riemann-Roch, n points $\{p_1,\ldots,p_n\}$ of C are collinear in $\mathbf{P}^3$ if and only if $h^0(\sum_1^n p_i) = n-1$ so that $n \leq 3$ by Clifford's theorem. Thus at most 3 points of C are collinear and hence $\gamma^{-1}(\gamma(p,q)) = (p,q)$ unless for some $r \in C$, $h^0(p+q+r) = 2$.

Thus, since C has at most two g_3^1's, γ is birational onto its image, and the multiple locus of γ is an analytic set $R \subseteq C^{(2)}$ consisting precisely of those pairs (p,q) such that for some $r \in C$, $|p+q+r|$ is a g_3^1 on C. Let $B = \gamma(R)$ be the image of the multiple locus. Then B = the set of lines $\ell \in \mathbb{G}r(1,3)$ such that $\#(\ell\cdot C) = 3$, where $(\ell\cdot C)$ is defined as the g.c.d. of the divisors $\{(H\cdot C)$ where H is a plane in $\mathbf{P}^3$ containing $\ell\}$.

Since $C = Q_2 \cap F_3$ is the complete intersection of a quadric and a cubic, Bezout's theorem implies $\ell \in B$ if and only if ℓ lies on the quadric Q.

Hence $B \subseteq \mathbb{G}r(1,3)$ parametrizes the lines on an irreducible quadric surface, so either $B \cong \mathbf{P}^1$ or $B \cong \mathbf{P}^1 \amalg \mathbf{P}^1$, according to whether Q has rank 3 or 4. Assume for concreteness, that $B \cong \mathbf{P}_a^1 \amalg \mathbf{P}_b^1$. Then $Q \cong \mathbf{P}_a^1 \times \mathbf{P}_b^1$ and the two projections to $\mathbf{P}^1$ restrict on C to the rational maps defined by the g_3^1's L_a and L_b on C.

Hence as ℓ runs through one copy of $\mathbf{P}^1$ the divisors $(\ell\cdot C)$ run through one of the linear systems g_3^1 on C. Now denote $\gamma^{-1}(\mathbf{P}_a^1) = C_a$, and $\gamma^{-1}(\mathbf{P}_b^1) = C_b$, so that $R = C_a \amalg C_b$. Now map $C \to C_a$ by $p \to D_a(p)-p$; where $D_a(p) \in |L_a|$ is the unique divisor in L_a containing p.

By choosing local sections of the map $\phi_{L_a}: C \to \mathbf{P}^1$, and using the extension theorem for holomorphic maps on smooth curves; the map $C \to C_a$ is an analytic bijection, hence finite algebraic and birational and the diagram below commutes:

$$\begin{array}{ccc} C & \longrightarrow & C_a \\ & {\scriptstyle \phi_{L_a}} \searrow \quad \swarrow {\scriptstyle \gamma} & \\ & \mathbf{P}^1 & \end{array}$$

The same holds for C_b, and the proposition is proved. We may state it as follows:

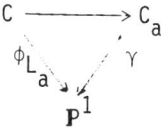

COROLLARY. A non-hyperelliptic genus 4 curve C and its g_3^1's are completely determined by restricting the canonical map ϕ of the variety $C^{(2)} \cong W_2$ to its multiple locus; i.e. the multiple locus itself is birational equivalent to as many copies of C as there are g_3^1's, and the restriction of ϕ to these copies of C gives the corresponding rational maps $\phi_{g_3^1}$.

<u>Questions</u>:

1. Are the components of the multiple locus actually isomorphic to C, or only birational?
2. ("Brill-Noether") Can one compute the multiple locus as a homology class on $C^{(2)}$ and show directly it is always $\neq 0$,

Now consider the hyperelliptic case. The g_2^1 is unique, so W_2 has one singular point, a triple double point x, the Abel map

$$\alpha : C^{(2)} \to W_2$$

is an isomorphism except over that point, and $\alpha^{-1}(x) \cong \mathbf{P}^1$. In the diagram

$$\begin{array}{ccc} C^{(2)} & \overset{\alpha_\star}{\dashrightarrow} & \mathbb{G}r(1,3) \\ & {\scriptstyle\alpha}\searrow \quad \nearrow{\scriptstyle\gamma} & \\ & W_2 & \end{array}$$

γ is undefined at x, and $\alpha_\star$ is undefined along $\alpha^{-1}(x) = |g_2^1|$. If $\phi_K : C \to \mathbf{P}^3$ is the canonical map of C, then $\phi_K(C) = \Gamma$ is a smooth rational twisted cubic, and ϕ_K is the map of degree two defined by the unique g_2^1 (followed by the cubic Veronese map). Thus $\alpha_\star(p,q) = \overline{\phi_K(P)\phi_K(q)}$ is the secant of Γ determined by the canonical images of p and q. ($\alpha_\star$ actually extends to a morphism on $C^{(2)}$). We will identify $C^{(2)} - |g_2^1|$ with $W_2 - \{x\}$ via α. Since ϕ_K has degree two, γ is no longer birational but a general secant line of Γ now has 4 preimages in $W_2 - \{x\}$. A tangent line $T_{\phi(p)}(\Gamma)$ has only two preimages under γ, (p,p) and (q,q) where $|p+q| = g_2^1$, $p \neq q$, ((p,q) is another preimage under the extended $\alpha_\star$). If $|2p| = g_2^1$, the tangent line $T_{\phi(p)}(\Gamma)$ has no preimages under γ (and only one (p,p) under the extension).

If $\overline{W_2}$ is the surface parametrizing secants to Γ, then $\overline{W_2}$ is birational to $(\mathbf{P}^1)^{(2)} \cong \mathbf{P}^2$, hence

$$\gamma : W_2 \dashrightarrow \overline{W_2} \subseteq \mathbb{G}r(1,3)$$

is a rational map to a rational surface, generically four to one, and missing ten points (the tangents to the ten branch points of the map $\phi_K: C \to \Gamma$). The family of tangent lines to Γ form a rational curve on $\overline{W_2}$, the unique curve in the branch locus of γ (as defined by Andreotti) which passes through all ten of the omitted points. The preimage of this curve is birational to C and γ restricts there to the map associated to the g^1_2.

Thus both the curve and the g^1_2 are recovered by restricting the Gauss map on W_2 to the preimage of a well-defined component of the branch locus.

Now recall that in the non-hyperelliptic case it was the g^1_3's that influenced the Gauss map on W_2. Here there is a one-parameter family of g^1_3's on the hyperelliptic C, all with a base point (by Castelnouvo's trick of computing the aritmethic genus of the image of $\phi_{g^1_2} \times \phi_{g^1_3}: C \to \mathbf{P}^1 \times \mathbf{P}^1$), and these also affect the Gauss map. In the rational surface $\overline{W_2}$ there are, of course, many rational curves over which γ is in general four to one with irreducible preimage, but for those ocurring as a pencil of secants of Γ through some fixed point, the preimage of γ is in general reducible with two components both birational to C, and γ restricts on both of them to the map $\phi_L, L = g^1_2$.

This family of copies of the map ϕ_L parametrized by C corresponds to the family of rational maps $\phi_{L_p}, L_p = p + g^1_2$, defined by the various g^1_3's on C.

<u>Question</u>:

Are these last pencils of secants characterized as the unique linear pencils in $\overline{W_2} \approx \mathbb{P}^2$ which have reducible preimages under γ?

§2. GENERAL PENCILS: g^1_d's IN ARBITRARY GENUS.

Assume for some $d \leq g$ that C has a g^1_d but no g^1_{d-1} and no g^2_{d+1}. Then the Abel map

$$\alpha: C^{(d-1)} \to W_{d-1} \subseteq J(C)$$

is an isomorphism, and we will recover the "universal" $\phi_{g^1_d}$ on C from the Gauss map (i.e. canonical map) on W_{d-1}. Taking the derivative of α we have the basic diagram:

$$\begin{array}{ccc} C^{(d-1)} & \xrightarrow{\alpha_*} & \mathbb{G}r(d-2,g-1) \\ \cup\! | & & \cup\! | \\ R & \xrightarrow{\alpha_*} & B \end{array}$$

where R = the "multiple locus" of α_* and $B = \alpha_*(R)$ is its image.

DEFINITION. If $f: X \to Y$ is a morphism, we define the "multiple locus" $R \subseteq X$ as follows (as a set): R = closure of $(\{x \in X\colon$ for some $z \in X$, $z \neq x$, and $f(z) = f(x)\})$.

LEMMA. With the hypotheses of this section,

$$\begin{aligned} R &= \{D \in W_{d-1} : \#(\overline{D}\cdot\Gamma) \geq d\} \\ &= \{D \in W_{d-1} : \#(\overline{D}\cdot\Gamma) = d\} \\ &= \{D \in W_{d-1} : \text{ for some } p \in C, \\ &\qquad |D + p| = \text{a } g^1_d\}. \end{aligned}$$

PROOF. By definition, $(\overline{D}\cdot\Gamma)$ = g.c.d. $\{(H\cdot\Gamma)\colon H$ is a hyperplane in $\mathbf{P}^{g-1}$ and $(H\cdot\Gamma) \geq D\}$, so we always have $D \leq (\overline{D}\cdot\Gamma)$. If $D \in W_{d-1}$, then the geometric Riemann-Roch theorem implies that $\overline{D}$ is a (d-2)-plane since there are assumed to be no g^1_{d-1}'s. Thus if

$$\#(\overline{D}\cdot\Gamma) \geq d+1$$

then $|(\overline{D}\cdot\Gamma)|$ would contain a g^2_{d+1}, again by geometric Riemann-Roch, contrary to hypothesis. Hence for any $D \in W_{d-1}$, $\#(\overline{D}\cdot\Gamma) \geq d$ if and only if $\#(\overline{D}\cdot\Gamma) = d$, which proves the equivalence of two of our conditions. Now let $D \in W_{d-1}$ be such that $|D+p|$ = a g^1_d. Then geometric Riemann-Roch implies $\overline{(D+p)}$ is a (d-2)-plane, and since $\overline{D}$ is always a (d-2)-plane, $\overline{D} = \overline{(D+p)}$. Therefore, since $D+p \leq (\overline{D}\cdot\Gamma)$, we get $\#(\overline{D}\cdot\Gamma) \geq d$. Now conversely, assume $D \in W_{d-1}$ and $\#(\overline{D}\cdot\Gamma) = d$. Then $D \leq (\overline{D}\cdot\Gamma)$ implies $(\overline{D}\cdot\Gamma) = D+p$, and $\overline{D} = (\overline{D+p})$, so again geometric Riemann-Roch implies $(\overline{D+p})$ = a g^1_d. Thus the 3 conditions in the lemma are equivalent, and since the first condition is closed they define a common closed set $S \subseteq W_{d-1}$. We show next that $S = R$. Since S is closed, to show $R \subseteq S$ it suffices to check that if $D, E \in W_{d-1}$, $D \neq E$ and $\overline{D} = \overline{E}$, that then $\#(\overline{D} \cdot \Gamma) \geq d$. But $D \leq (\overline{D} \cdot \Gamma)$ and $E \leq (\overline{E} \cdot \Gamma)$, so if $\overline{D} = \overline{E}$ and $\#(\overline{D} \cdot \Gamma) = d-1$, then $D = (\overline{D} \cdot \Gamma) = (\overline{E} \cdot \Gamma) = E$, a contradiction. To show $S \subseteq R$, assume $D \in W_{d-1}$ and $\#(\overline{D} \cdot \Gamma) = d$. Then $|(\overline{D} \cdot \Gamma)| =$ a g^1_d with no fix points (there is no g^1_{d-1}, by assumption). Thus for the general divisor in $|(\overline{D} \cdot \Gamma)|$ every subset of cardinality $d-1$ gives a point of R, and our divisor D is in the closure of these points of R. The lemma is proved.

COROLLARY.

1) $B = \alpha_*(R) = \{$the set of d-secant (d-2)-planes in $\mathbf{P}^{g-1}\}$
$= \{$linear spaces of form $\overline{D}$ for some D in some $g^1_d\}$
2) If α now denotes the Abel map on $C^{(d)}$, $\alpha : C^{(d)} \to W_d$, and if $W_d \supseteq W^1_d = \{$the set of g^1_d's on $C\}$, then the projectivized derivative of α gives a holomorphic bijection (and thus a homeomorphism in the complex topology)

$$\alpha_* : \alpha^{-1}(W^1_d) \to B$$

where $\alpha_*(D) = \overline{D}$.

LEMMA. B is fibered in precisely one way by a family of disjoint analytic curves homeomorphic to $\mathbf{P}^1$; precisely, the only such curves on B are those of form:

$$\{\overline{D}: \text{ for all } D \text{ in some fixed } g^1_d \text{ on } C\}.$$

PROOF. It is sufficient by part 2) of the previous corollary to prove the same statement for the set

$$\alpha^{-1}(W^1_d) = \{\text{all divisors } D \text{ on } C \text{ such that } |D| \text{ is a } g^1_d\}.$$

Let $X \subseteq C^{(d)}$ be any analytic curve homeomorphic to $\mathbf{P}^1$. Then Abel's map, restricted to X, factors through the universal covering space of $J(C)$:

$$\begin{array}{ccc} & & \mathbb{C}^g \\ & \overset{\tilde{\alpha}}{\nearrow} & \downarrow \\ X & \xrightarrow{\alpha} & W_d \subseteq J(C) \end{array}$$

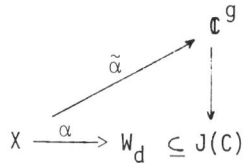

Since X is compact, and $\tilde{\alpha}$ is analytic, the maximum modulus principle implies that $\tilde{\alpha}(X)$ is a single point, so that $\alpha(X)$ is also a single point. Then by the converse of Abel's theorem, $X = |D|$ = the set of divisors belonging to some g^1_d on C.

Now we state a theorem describing the Gauss map on W_{d-1} for a generic d-gonal curve.

THEOREM. If C has a g^1_d but no g^1_{d-1} and no g^2_{d+1}, for some d with $d \le g$ = genus (C), if $\phi: W_{d-1} \to \mathbb{G}r(d-2,g-1)$ is the canonical map on W_{d-1}, if R is the multiple locus of ϕ, $B = \phi(R)$, then the restriction $\phi: R \to B$ is a "universal" $\phi_{g^1_d}$ for C. Precisely, B is uniquely a union of a family of disjoint algebraic curves homeomorphic to $\mathbf{P}^1$; these are precisely the curves of the form $\alpha_*(|g^1_d|)$ where $|g^1_d| \subseteq C^{(d)}$ is a pencil of linearly equivalent divisors; for any one of these curves $X_L \subseteq B$, $L = g^1_d$, the inverse image $\phi^{-1}(X_L) \subseteq R$ is an algebraic curve bijectively birational to C; after normalizing, the Gauss map of W_{d-1}, restricted to $\phi^{-1}(X_L)$, induces the rational map ϕ_L determined by the linear series $L = g^1_d$, i.e. the following diagram commutes:

$$\begin{array}{ccc} C & \xrightarrow{\ \phi_L\ } & \mathbf{P}^1 \\ {\scriptstyle \nu}\downarrow & & \downarrow{\scriptstyle \nu} \\ \phi^{-1}(X_L) & \xrightarrow{\ \phi\ } & X_L \end{array}$$

in which ν denotes normalization.

PROOF. We have proved all but the last two statements. Since $\phi: R \to B$ is a proper map with finite fibers it is a finite map, $\phi^{-1}(X_L)$ is a curve, and the map

$$f: C \to \phi^{-1}(X_L)$$

$$p \mapsto (\overline{D(p)} \cdot C) - \{p\}$$

is a holomorphic bijection, hence it is the normalization map, where $D(p) \in g^1_d = L$ is the unique divisor of L containing p. The normalization diagram

$$\begin{array}{ccc} C & \xrightarrow{\phi_L} & |L| \subseteq C^{(d)} \\ {\scriptstyle f=\nu}\downarrow & & \downarrow{\scriptstyle \alpha_* = \nu} \\ \phi^{-1}(X_L) & \xrightarrow{\phi} & X_L \subseteq B \end{array}$$

is now seen to commute.

Question:

The geometric Riemann-Roch theorem implies that if the multiple locus of $\phi: W_{d-1} \to \mathbb{G}r(d-2,g-1)$ is non-empty, then there exists a g^1_d. Can one compute directly that when the Brill-Noether number $\rho = g-2(g-d+1)$ is positive, that then this multiple locus carries a non-zero homology class?

§3. A NON-GENERIC EXAMPLE.

We want to give one slightly more complicated example in which the hypotheses of the previous theorem are violated and the structure of the Gauss map becomes surprisingly richer and even more interesting. Let C now be a trigonal, non-hyperelliptic curve of genus 5; i.e. there is a g^1_3 but no g^1_2. Then $K - g^1_3 = g^2_5$ so in fact both hypotheses fail to hold.

The g^1_3 corresponds to a unique copy of $\mathbf{P}^1 \cong |g^1_3|$ in $C^{(3)}$ and the canonical map

$$\phi: C^{(3)} \dashrightarrow \mathbb{G}r(2,4)$$

is defined on the complement $C^{(3)} - |g^1_3|$. Since we are looking for the multiple locus, we seek distinct triples of points on C that span the same 2-plane in $\mathbb{P}^4$. We may identify C with its canonical model $\Gamma \subseteq \mathbf{P}^4$, and so we want 2-planes in $\mathbf{P}^4$ which are at least 4-secant (with respect to Γ), and equivalently triples of points on C that belong to divisors of g^1_4's. There are now two kinds of g^1_4's on C, those of form $p + g^1_3$, with a fixed point, and those of form $K - (p + g^1_3) = g^2_5 - q$, generically without a fixed point, according to Andreotti and Mayer [2,p.212]. Those are precisely the two types that were prohibited by the hypotheses of the previous theorem, so there are no "generic" g^1_4's in the present example. A 2-plane cannot be 6-secant (or more) since it would then cut out a divisor of a g^3_6 and C would be hyperelliptic, contrary to hypothesis. Thus $B \subseteq \mathbb{G}r(2,4)$ consists of 4-secant and 5-secant 2-planes. The 5-secant 2-planes are spanned by the divisors of the unique $g^2_5 = K - g^1_3$, hence are parametrized by $\mathbb{P}^2$. The 4-secant 2-planes are spanned by divisors of the linear series $(g_3^1 + p)$ and are thus parametrized by a $\mathbf{P}^1$- bundle over C. B is thus a reducible surface which is connected as follows: if we denote $|g_5^2 - g_3^1| = |g_2^0|$ by u+v, then in B the three pencils of 2-planes corresponding to the 3 sets of divisors $|g_3^1 + u|$, $|g_3^1 + v|$, {all those $D \in |g_5^2|$ such that $D \geq u+v$} , are all the same. To see this and to follow all the rest of this section, it is best to keep in mind the plane quintic model of C, a quintic Δ with one double point:

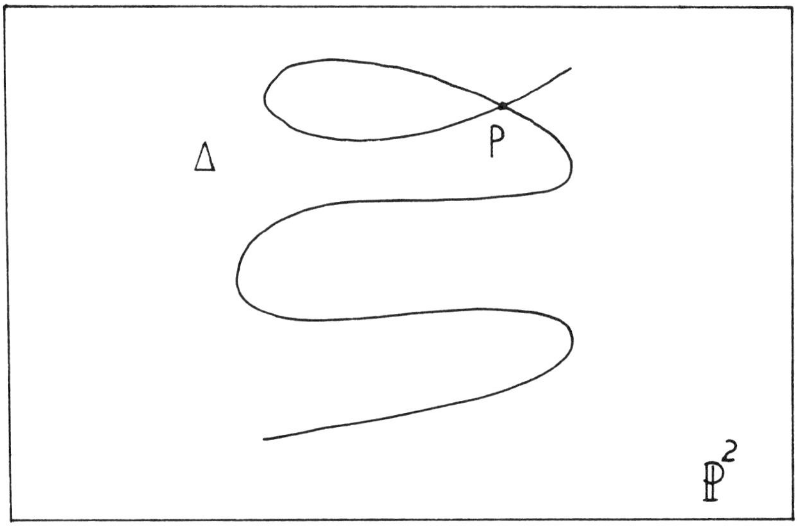

the double point p arising from the identification of the points u and v. The divisors of the g_3^1 are those cut by the lines in this plane, residually to the double point; the divisors of the g_5^2 are those cut by arbitrary lines in the plane; the divisors of the fix-point-free g_4^1's are cut residually by lines which pass through some base point on Δ (other than the double point p).

Now what is the structure of ϕ over B? Over the component of 4-secant 2-planes ϕ has degree three (since it is undefined at a divisor in the g_3^1). As in our earlier examples, ϕ restricts over this component to the union of the family of rational maps $\{\phi_L = x + g_3^1,\ x \in \Delta - \{p\}\}$. Over the component of 5-secant 2-planes, which is parametrized by $(\mathbf{P}^2)^* \cong |g_5^2| \cong$ the dual of the ambient plane containing the g_5^2 - model Δ of C, ϕ is a little more complicated. Each 2-plane in this component of B cuts five points of Γ, any three of which span the 2-plane (except the triples belonging to the g_3^1, and which occur in only one pencil of 2-planes contained in $(\mathbf{P}^2)^*$). Thus ϕ has degree $\binom{5}{3} = 10$ over this component of B which reflects the existence of the g_4^1's of form $|g_5^2 - x|$. In fact ϕ extends to a proper map over the whole component $\cong (\mathbf{P}^2)^*$ in B, by mapping any triple of points in a divisor

$|g_5^2 - x|$ first to the unique divisor of $|g_5^2|$ containing it and then to the 2-plane that divisor spans. Then we see that the branch locus of ϕ over this component is exactly Δ^*, the (irreducible) dual of the plane quintic model of C. Thus to each non-singular point $x \in \Delta$ we can associate both a fix-point-free g_4^1, namely $|g_5^2 - x|$ and a tangent line ℓ_x to Δ^* at the "point" of Δ^* represented by $T_x(\Delta)$. Recall that a tangent line $\ell_x \subseteq (\mathbf{P}^2)^*$ to Δ^* at $T_x(\Delta)$ is given by the pencil of lines in $\mathbf{P}^2$ which contain the point $x \in \Delta$. Let us restrict ϕ to the locus over ℓ_x. Since we can distinguish triples of point which contain x from those which do not, the inverse image $\phi^{-1}(\ell_x)$ is reducible. The component of triples not containing x gives a birational copy of C on which ϕ restricts to ϕ_L, $L = |g_5^2 - x|$, in the same way as in the examples of the earlier sections. The choices of triples which contain x, on the other hand, are equivalent to choices of pairs of points in the divisors of the linear series $g_4^1 = |g_5^2 - x|$. Hence ϕ has degree 6 on this component. Moreover disjoint pairs in the same divisor can be exchanged, so this component of $\phi^{-1}(\ell_x)$ is a curve with an involution and a symmetric g_6^1, and therefore the quotient curve under the involution has a g_3^1.

According to S. Recillas [5], this is in fact the double cover of a trigonal genus 6 curve associated to the pair $(C, |g_5^2 - x|)$, and whose Prym variety is the Jacobian variety J(C) of our original curve! For completeness, consider ℓ_p = pencil of lines in $\mathbf{P}^2$ passing through p, the double point of Δ. (This is the tangent line to two points of Δ^*). Restricting ϕ to it gives two g_4^1's of form $|g_5^2 - x|$, i.e. $|g_5^2 - u| = g_3^1 + v$, and $|g_5^2 - v| = g_3^1 + u$. (Recall that $u+v = |g_5^2 - g_3^1|$ are the two points of C over the double point of Δ. The preimage $\phi^{-1}(\ell_p)$ has four components: one copy of $|g_3^1| \cong \mathbf{P}^1$, by choosing triples residual to the double point, where ϕ is essentially the identity; one copy of C from choosing triples containing u, where ϕ equals $\phi_{g_3^1+v} = \phi_{g_3^1}$; one copy of C from choosing triples contianing v, where ϕ equals $\phi_{g_3^1+u} = \phi_{g_3^1}$; and one copy of C from choosing triples containing u+v, where $\phi = \phi_{g_3^1}$. (We remark that as $x \to p$, the degeneration $\phi^{-1}(\ell_x) \to \phi^{-1}(\ell_p)$ is quite pretty and illuminating: it exhibits the "Wirtinger" degeneration from Prym theory of an étale irreducible double cover into an étale reducible double cover, and also the degeneration of a degree four map of $C \to \mathbf{P}^1$ into a map $C \cup \mathbf{P}^1 \to \mathbf{P}^1$ which has degree three on C and degree one on $\mathbf{P}^1$, a basic phenomenon in understanding specialization of linear series on curves).

SUMMARY. If C is a plane quintic of genus five, (therefore trigonal and non-hyperelliptic), the canonical map of $C^{(3)}$ determines, in a direct geometric way, not only the rational maps arising from the g_5^2, the g_3^1, and all the g_4^1's, but also those associated to the symmetric g_6^1's on curves whose Prym variety is $J(C)$!

BIBLIOGRAPHY

[1] A. Andreotti, On a theorem of Torelli, Am. J. Math., 80(1958).

[2] A. Andreotti, and A. Mayer, On period relations for abelian integrals..., Ann. della Scuola Norm. Sup. Pisa, vol. XXI, 1967, 189-238.

[3] P. Griffiths, and J. Harris, Principles of Algebraic Geometry, New York, John Wiley and Sons, 1978.

[4] D. Mumford, Curves and their Jacobians, Ann Arbor, The University of Michigan Press, 1975.

[5] S. Recillas, Jacobians of curves with a g_4^1 are Prym varieties of trigonal curves, Bol. Soc. Mat. Mexicana, 19(1974), 9-13.

Roy Smith
Department of Mathematics
University of Georgia
Athens, Georgia, U.S.A.

Horacio Tapia-Recillas
Departamento de Matemáticas
CINVESTAV, IPN and UAM, I.
Apdo. Postal 14-740
México 07000 D.F., México

Contemporary Mathematics
Volume 58, Part I, 1986

Some applications of algebraic geometry to systems of partial differential equations and to approximation theory

Peter F. Stiller

ABSTRACT. We shall discuss the application of some results in the theory of vector bundles/reflexive sheaves on complex projective spaces to systems of constant coefficient partial differential equations and to certain local problems in the theory of multivariate spline approximation.

An introduction to the problems.

At the time I was introduced to the problems discussed below, I was investigating the various ways that existed for constructing holomorphic vector bundles on $\mathbb{P}^n_{\mathbb{C}}$ (particularly low ranks < n). It quickly became apparent that several questions in approximation theory could be interpreted as questions about the cohomology of a suitably constructed vector bundle. This led to my subsequent interest in these problems.

The ideas evolved as the result of a number of discussions with Dr. Charles Micchelli, IBM research, and to him must go the credit for introducing the author to these problems from the point of view of approximation theory. I have omitted some of the proofs (a few are straightforward and the others will appear in detail elsewhere).

A special note of thanks must go to the organizers of the Lefschetz Centennial Conference for the work they did in making it such a memorable experience.

We begin with some notation.

Let $P^{k,n}_{\mathbb{R}}$ $(P^{k,n}_{\mathbb{C}})$ denote the space of homogeneous polynomials of degree k in n variables with coefficients in the real numbers $\mathbb{R}$ (complex numbers $\mathbb{C}$). We have $\dim_{\mathbb{R}} P^{k,n}_{\mathbb{R}} = \binom{n+k-1}{n-1} = \binom{n+k-1}{k}$ and $\dim_{\mathbb{C}} P^{k,n}_{\mathbb{C}} = \binom{n+k-1}{n-1} = \binom{n+k-1}{k}$.

1980 Mathematics Subject Classification. 14F05, 32L10, 35Exx, 41A15

Let $p_1,\dots,p_m$ be homogeneous polynomials of degree $d_1,\dots,d_m$ respectively in $\mathbb{R}[x_1,\dots,x_n]$, i.e.

$$p_i \in P_{\mathbb{R}}^{d_i,n},$$

and let $p_i(\partial)$ denote the corresponding linear homogeneous partial differential operator with constant coefficients given by

$$p_i(\partial) = p_i(\frac{\partial}{\partial x_1},\dots,\frac{\partial}{\partial x_n}).$$

We denote by D the common null space of the operators $p_i(\partial)$ in the space of real-valued infinitely differentiable functions on $\mathbb{R}^n$:

$$D = \{f \in \mathcal{C}^{\infty}_{\mathbb{R}^n}: \ p_i(\partial)f = 0 \ \text{ for } \ i = 1,\dots,m\}.$$

We then have the following results which were introduced to the author by Dr. Charles Micchelli, IBM Research Center:

Proposition 1 (Micchelli): $\dim_{\mathbb{R}} D < \infty$ if and only if $\{z \in \mathbb{C}^n - \{(0,\dots,0)\}$ s.t. $p_i(z) = 0 \ \ i = 1,\dots,m\} = \emptyset$. In other words, D is finite dimensional if and only if the hypersurfaces $p_i = 0$ in $\mathbb{P}^{n-1}_{\mathbb{C}}$ have empty intersection (requiring in particular at least as many equations as unknowns, i.e. $m \geq n$). □

The proof involves a straightforward application of the Nullstellensatz, and has two interesting facets. First, the proof shows in the finite dimensional case, $\dim_{\mathbb{R}} D < \infty$, that the null space D consists entirely of polynomials, and secondly in the infinite dimensional case, that there are polynomial solutions of every degree.

Proposition 2 (Micchelli): If $f \in D$ and $f = \sum_{r=0}^{N} f_r$, $f_r \in \mathcal{C}^{\infty}_{\mathbb{R}^n}$ where each f_r is homogeneous of degree r then $f_r \in D$. □

We define:

$$D_k = \{p \in P^{k,n}_{\mathbb{R}}: \ p_i(\partial)p = 0 \ \ i = 1,\dots,m\}.$$

Thus in the case when D is finite dimensional we have

$$D = \bigoplus_{k=0}^{\infty} D_k$$

where $D_k = 0$ for k sufficiently large, and in the infinite dimensional case we have

$$\bigoplus_{k=0}^{\infty} D_k = D^{poly} \subset D$$

where D^{poly} are the polynomial solutions to our system of partial differential equations

$$p_i(\partial)f = 0 \qquad i = 1,\ldots,m.$$

We now arrive at a first problem:

Problem 1: Find $\dim_{\mathbb{R}} D_k$. (Particularly useful in the case $\dim_{\mathbb{R}} D < \infty$ is to find the least integer k_0 such that $\dim_{\mathbb{R}} D_k = 0$ for $k > k_0$.)

To formulate a second problem, we again take $p_1,\ldots,p_m$ with $p_i \in P_{\mathbb{R}}^{d_i,n}$ homogeneous polynomials of degree d_i in n variables and consider the space V_k of polynomial relations of total degree k:

$$V_k = \{(q_1,\ldots,q_m) \quad q_i \in P_{\mathbb{R}}^{k-d_i,n} : \sum_{i=1}^{m} q_i p_i = 0\}.$$

Here it is understood that $P_{\mathbb{R}}^{k-d_i,n} = 0$ if $d_i > k$.

The spaces V_k admit an interesting interpretation in approximation theory. They arise in the context of conformality conditions and in the construction of multivariate splines (see Chui and Wang [1] and Stiller [10]). To make this connection clear we consider a simple low dimensional local situation. In $\mathbb{R}^2$, take a number of polynomial arcs $A_1,\ldots,A_N$ emanating from the origin so that each A_i is a piece of the locus $p_i(x,y) = 0$, $p_i(0,0) = 0$, for some irreducible polynomial $p_i(x,y) \in \mathbb{R}[x,y]$ of degree $d_i \geq 1$. We focus on a small neighborhood about the origin--small enough so that the arcs do not intersect except at the origin and so that they are smooth except perhaps at the origin.

Let $s(x,y)$ denote a spline function of order μ and degree k near the origin for this particular local triangulation. Assuming the arcs are numbered consecutively around the origin, denote by R_i the region between arcs A_i and A_{i+1} for $i = 1,\ldots,N$ where $A_{N+1} = A_1$. The spline function $s(x,y)$ is given by a polynomial $f_i(x,y)$ on R_i of degree $\leq k$ with f_i and f_{i+1} agreeing to order μ on A_{i+1}. This means that $f_{i+1} - f_i$ must lie in the $(\mu + 1)$st power of the ideal in $\mathbb{R}[x,y]$ defining the irreducible curve of which A_{i+1} is a part. In the case we are dealing with this means

$$f_{i+1} - f_i = q_{i+1} p_{i+1}^{\mu+1}$$

for some polynomial $q_{i+1} \in \mathbb{R}[x,y]$ of degree $\leq k - (\mu + 1)d_i$. If we add these equations we get the conformality relation

$$0 = \sum_{i=1}^{N} q_i p_i^{\mu+1}.$$

This is analogous to the space V_k.

<u>Problem 2</u>: Find $\dim_{\mathbb{R}} V_k$. (Clearly $\dim_{\mathbb{R}} V_k$ increases with k, and an interesting question is to understand this growth as a function of k.)

In answering these questions, we shall show below that V_k can be interpreted cohomologically as $H^0(\mathbb{P}_{\mathbb{C}}^{n-1}, E(k))$ and D_k as $H^1(\mathbb{P}_{\mathbb{C}}^{n-1}, E(k))$ for a suitable reflexive sheaf E on $\mathbb{P}_{\mathbb{C}}^{n-1}$. This permits the use of vector bundle techniques in the analysis of these systems of partial differential equations and in the construction of multivariate splines. This will be discussed below. For now, we recall another result due to Micchelli which relates D_k to V_k.

Consider the linear map

$$\begin{array}{ccc} P_{\mathbb{R}}^{k,n} & \xrightarrow{T} & \bigoplus_{i=1}^{m} P_{\mathbb{R}}^{k-d_i,n} \\ p & \longmapsto & (\ldots, p_i(\partial)p, \ldots) \end{array}$$

given by our system of differential operators $p_i(\partial)$. The null space (kernel) of T is D_k. We make $P_{\mathbb{C}}^{k,n}$ into a Hilbert space by defining $\langle p,q\rangle = p(\partial)\bar{q}$. We then have:

<u>Proposition 3</u> (Micchelli): The null space (kernel) of the adjoint T^* can be identified with V_k. It follows that

$$\dim_{\mathbb{R}} V_k - \dim_{\mathbb{R}} D_k = \sum_{i=1}^{m} \binom{n+k-d_i-1}{n-1}_+ - \binom{n+k-1}{n-1}$$

(where $+$ means the value is 0 if $k - d_i < 0$). □

We remark that the map T decomposes $P_{\mathbb{C}}^{k,n}$ into an orthogonal direct sum as

$$P_{\mathbb{C}}^{k,n} \cong (D_k \otimes_{\mathbb{R}} \mathbb{C}) \oplus I_k$$

where $I = \bigoplus_k I_k$ is the homogeneous ideal in $\mathbb{C}[x_1,\ldots,x_n]$ generated by $p_1,\ldots,p_m$.

In the case when D is finite dimensional, that is when the p_i have no common solution in $\mathbb{P}^{n-1}_{\mathbb{C}}$, we obviously have $D_k = 0$ for k sufficiently large. Thus

Corollary 4: $\dim_{\mathbb{R}} V_k = \sum_{i=1}^{m} \binom{n+k-d_i-1}{n-1}_+ - \binom{n+k-1}{n-1}$ for k sufficiently large, when the p_i have no common solution in $\mathbb{P}^{n-1}_{\mathbb{C}}$ (see also Stiller [10], Theorem 3.1). □

On the other hand in the case D is infinite dimensional, let $X \subset \mathbb{P}^{n-1}_{\mathbb{C}}$ be the non-empty closed subscheme defined by the coherent sheaf of ideals $\mathcal{I}_X$ associated to the homogeneous ideal generated by the p_i in $\mathbb{C}[x_1..x_n]$. Let $\mathcal{O}_X$ be the structure sheaf of X, then we shall show,

Corollary 5: $\dim_{\mathbb{R}} D_k = \dim_{\mathbb{C}} H^0(X,\mathcal{O}_X(k))$ for k sufficiently large. (Of course this is also valid in the case where D is finite dimensional because $X = \emptyset$ and $\dim_{\mathbb{C}} H^0(X,\mathcal{O}_X(k)) = 0$.) □

As a special case of Corollary 2.5, when X is supported on a finite set of points, we will have $\dim_{\mathbb{R}} D_k$ independent of k for k sufficiently large, i.e. constant, because in that case $\mathcal{O}_X(k) \cong \mathcal{O}_X$. In the general case, we have:

Proposition 6: $\dim_{\mathbb{R}} D_k$ for k large can be given by a polynomial in ϕ with rational coefficients and degree equal to the dimension of X.

Proof: We have $\dim_{\mathbb{R}} D_k = \dim_{\mathbb{C}} H^0(X, \mathcal{O}_X(k))$ for k large, so the polynomial will be the Hilbert polynomial for X (see Hartshorne [4], Chap. 1, §7).

The crucial step in determining $\dim_{\mathbb{R}} D_k$, $\dim_{\mathbb{R}} V_k$, and $\dim_{\mathbb{R}} D$, where D is the common null space for the system of partial differential operators under consideration, is to reinterpret the problem in the language of vector bundles. This we do below.

Translation to Vector Bundles/Reflexive Sheaves.

As before, let $p_1,\dots,p_m$ be homogeneous polynomials in $\mathbb{R}[x_1 \dots x_n]$, $n \geq 2$, and say p_i has degree $d_i \geq 1$ so that

$$p_i \in P^{d_i,n}_{\mathbb{R}}.$$

We identify $P^{d_i,n}_{\mathbb{C}}$ with $H^0(\mathbb{P}^{n-1}_{\mathbb{C}}, \mathcal{O}_{\mathbb{P}^{n-1}_{\mathbb{C}}}(d_i))$, the space of global sections of the line bundle $\mathcal{O}_{\mathbb{P}^{n-1}_{\mathbb{C}}}(d_i)$ over complex projective $n-1$ space $\mathbb{P}^{n-1}_{\mathbb{C}}$. Viewed in this way each p_i gives rise to an injection

$$0 \to \mathcal{O}_{\mathbb{P}^{n-1}_{\mathbb{C}}}(-d_i) \to \mathcal{O}_{\mathbb{P}^{n-1}_{\mathbb{C}}}.$$

We consider the direct sum

$$\bigoplus_{i=1}^{N} \mathcal{O}_{\mathbb{P}^{n-1}_{\mathbb{C}}}(-d_i) \to \mathcal{O}_{\mathbb{P}^{n-1}_{\mathbb{C}}}.$$

The image of this map is a coherent sheaf of ideals $\mathcal{O}_X$ for a closed subscheme $X \subset \mathbb{P}^{n-1}_{\mathbb{C}}$ which is supported on the intersection of the hypersurfaces $p_i = 0$ in $\mathbb{P}^{n-1}_{\mathbb{C}}$. In fact, $\mathcal{I}_X$ is the coherent sheaf associated to the homogeneous ideal generated by the p_i in $\mathbb{C}[x_1..x_n]$. Note that $\mathcal{I}_X = \mathcal{O}_{\mathbb{P}^{n-1}_{\mathbb{C}}}$ when the p_i have no common solution in $\mathbb{P}^{n-1}_{\mathbb{C}}$ -- that is when D, the space of $\mathcal{C}^{\infty}$ solutions f to our system of partial differential equations, $p_i(\partial)f = 0$ $i = 1,\dots,m$, is finite dimensional.

We consider the sequence

$$0 \to K \to \bigoplus_{i=1}^{m} \mathcal{O}_{\mathbb{P}^{n-1}_{\mathbb{C}}}(-d_i) \to \mathcal{I}_X \to 0$$

where K is the kernel (a coherent sheaf on $\mathbb{P}^{n-1}_{\mathbb{C}}$) and all the sequences we get after tensoring with the line bundles $\mathcal{O}_{\mathbb{P}^{n-1}_{\mathbb{C}}}(k)$ $k \in \mathbb{Z}$:

$$0 \to K(k) \to \bigoplus_{i=1}^{m} \mathcal{O}_{\mathbb{P}^{n-1}_{\mathbb{C}}}(k-d_i) \to \mathcal{I}_X(k) \to 0.$$

Proposition 7: $\dim_{\mathbb{C}} H^0(\mathbb{P}^{n-1}_{\mathbb{C}}, K(k)) = \dim_{\mathbb{R}} V_k$. □

Proposition 8: K is a reflexive sheaf on $\mathbb{P}^{n-1}_{\mathbb{C}}$ and is in fact locally free (vector bundle) in the case $n = 2,3$ or in the case $n \geq 4$ when the p_i have no common solution in $\mathbb{P}^{n-1}_{\mathbb{C}}$.

Proof: To show K is reflexive it is enough to show it is torsion free and normal. Torsion free is obvious, and normality results from the fact that $\mathcal{I}_X$ is torsion free (see Okonek, Schneider, Spindler [8], 1.1.12 Lemma p. 150 and 1.1.6 Lemma p. 158). Now a reflexive sheaf has its singularity set in codimension 3, so that when $n = 2$, $\mathbb{P}^1_{\mathbb{C}}$, or $n = 3$, $\mathbb{P}^2_{\mathbb{C}}$, it must be locally free. Finally if the p_i have no common solution in $\mathbb{P}^{n-1}_{\mathbb{C}}$ then K is the kernel of a surjective map of locally free sheaves (vector bundles)

$$0 \to K \to \bigoplus_{i=1}^{m} \mathcal{O}(-d_i) \to \mathcal{O} \to 0$$

and hence is locally free. □

<u>Propostion 9</u>: For k sufficiently large

$$\dim_{\mathbb{C}} H^0(\mathbb{P}^{n-1}_{\mathbb{C}}, K(k)) = \sum_{i=1}^{m} \binom{n+k-d_i-1}{n-1}_+ - \binom{n+k-1}{n-1} + \dim_{\mathbb{C}} H^0(\mathcal{O}_X(k))$$

where the + indicates that we take $\binom{n+k-d_i-1}{n-1} = 0$ if $k - d_i < 0$, and $\mathcal{O}_X$, the cokernel in

$$0 \to \mathcal{I}_X \to \mathcal{O}_{\mathbb{P}^{n-1}_{\mathbb{C}}} \to \mathcal{O}_X \to 0 ,$$

is the structure sheaf of the closed subscheme X in $\mathbb{P}^{n-1}_{\mathbb{C}}$. □

Combining this proposition with Proposition 3 and 7 gives Corollary 4 and Corollary 5.

We come now to the main result that allows us to translate the two problems above--namely $\dim_{\mathbb{R}} V_k = ?$ and $\dim_{\mathbb{R}} D_k = ?$, into problems about vector bundles/reflexive sheaves on $\mathbb{P}^{n-1}_{\mathbb{C}}$.

<u>Theorem 10</u>: Let $p_1,\dots,p_m \in \mathbb{R}[x_1..x_n]$ be homogeneous polynomials in n-variables of degrees $d_1,\dots,d_m$ respectively. Consider the corresponding partial differential operators $p_i(\partial) \in \mathbb{R}[\frac{\partial}{\partial x_1} \dots \frac{\partial}{\partial x_n}]$ and let D be the space of $\mathcal{C}^\infty_{\mathbb{R}^n}$ solutions to this system of partial differential equations and let D_k be the homogeneous polynomials of degree k in D:

$$D = \{f \in \mathcal{C}^\infty_{\mathbb{R}^n} \text{ s.t. } p_i(\partial)f = 0 \quad i = 1,\dots,m\}$$

$$D_k = D \cap P^{k,n}_{\mathbb{R}}.$$

If $n \geq 3$ and the p_i have no common solution in $\mathbb{P}^{n-1}_{\mathbb{C}}$ then we have for all k that

$$\dim_{\mathbb{R}} D_k = \dim_{\mathbb{C}} H^1(\mathbb{P}_{\mathbb{C}}^{n-1}, K(k))$$

and

$$\dim_{\mathbb{R}} D = \sum_{k\geq 0} \dim_{\mathbb{R}} D_k = \sum_{k\in\mathbb{Z}} \dim_{\mathbb{C}} H^1(\mathbb{P}_{\mathbb{C}}^{n-1}, K(k))$$

where $K(k)$ is the sheaf constructed above. (We should remark that $\bigoplus_{k\in\mathbb{Z}} H^1(\mathbb{P}_{\mathbb{C}}^{n-1}, K(k))$ can be viewed as a finite-length graded $\mathbb{C}[x_1..x_n]$-module and that this module appears in the vector bundle literature; see for example Hartshore [3].)

<u>Proof</u>: On $\mathbb{P}_{\mathbb{C}}^{n-1}$ $n \geq 3$ H^1 of any line bundle is zero. Thus for all $k \in \mathbb{Z}$ we have

$$0 \to H^0(K(k)) \to H^0(\bigoplus_{i=1}^{m} \mathcal{O}_{\mathbb{P}_{\mathbb{C}}^{n-1}}(k - d_i)) \to H^0(\mathcal{O}_{\mathbb{P}_{\mathbb{C}}^{n-1}}(k)) \to H^1(K(k)) \to 0$$

and

$$\dim_{\mathbb{C}} H^0(K(k)) - \dim_{\mathbb{C}} H^1(K(k)) = \sum_{i=1}^{m} \binom{n+k-d_i-1}{n-1}_+ - \binom{n+k-1}{n-1}.$$

We have seen that $\dim_{\mathbb{C}} H^0(K(k)) = \dim_{\mathbb{R}} V_k$, and this coupled with Proposition 3 gives

$$\dim_{\mathbb{C}} H^1(K(k)) = \dim_{\mathbb{R}} D_k$$

and the result follows. □

Notice that the case $n = 2$ is not covered in the theorem, but in that case the sheaf K is locally free on $\mathbb{P}_{\mathbb{C}}^1$ and so by a result of Grothendieck [2] isomorphic to a sum of line bundles $\mathcal{O}_{\mathbb{P}_{\mathbb{C}}^1}(a_i)$. Hence $\dim_{\mathbb{R}} V_k$ can be determined directly and from that $\dim_{\mathbb{R}} D_k$ and $\dim_{\mathbb{R}} D$. This is done using the results in Shatz [9]. We have only stated the conclusion below. The details will appear elsewhere.

<u>Systems of constant coefficient partial differential equations in two variables.</u>

Let $p_1,\ldots,p_m \in \mathbb{R}[x,y]$ be homogeneous polynomials of degree $d_1,\ldots,d_m$ respectively,

$$p_i \in P_{\mathbb{R}}^{d_i,2}, \quad d_i \geq 1.$$

For convenience, we order the p_i so that $1 \leq d_1 \leq d_2 \leq \dots \leq d_m$. Note that if the p_i have a greatest common factor $p \in P_{\mathbb{R}}^{d,2}$ of degree $d \geq 1$ then

$$\dim_{\mathbb{R}} D = \infty,$$

$$\dim_{\mathbb{R}} D_k = d \quad \text{for } k \text{ sufficiently large},$$

(because the scheme $X \subset \mathbb{P}^1_{\mathbb{C}}$ consists of d points if we count with multiplicity) and we can determine $\dim_{\mathbb{R}} V_k$ by considering the analogous problem for $\tilde{p}_i = \frac{p_i}{p} \in P_{\mathbb{R}}^{d_i-d,2}$. We will have

$$V_k = \tilde{V}_{k-d}$$

where $\tilde{V}_{k-d} = \{(q_1,\dots,q_m)\ \ q_i \in P_{\mathbb{R}}^{k-d_i} : \sum_{i=1}^{m} q_i\tilde{p}_i = 0\}$. In fact, we also have in this case that

$$\dim_{\mathbb{R}} D_k = \dim_{\mathbb{R}} \tilde{D}_{k-d} + d \quad \text{for } k \geq d$$

$$\dim_{\mathbb{R}} D_k = k + 1 \qquad\qquad \text{for } 0 \leq k < d.$$

Thus we can always reduce the case where the p_i have a common factor to one where they don't, and so from now on we shall assume $p_1,\dots,p_m$ have no common factor. In this case we have for every k the exact sequence of vector bundles (see above):

$$0 \to K(k) \to \bigoplus_{i=1}^{m} \mathcal{O}_{\mathbb{P}^1_{\mathbb{C}}}(k - d_i) \to \mathcal{O}_{\mathbb{P}^1_{\mathbb{C}}}(k) \to 0,$$

where $H^0(\mathbb{P}^1_{\mathbb{C}},K(k))$ can be identified with $V_k \otimes_{\mathbb{R}} \mathbb{C}$ so that

$$\dim_{\mathbb{C}} H^0(\mathbb{P}^1_{\mathbb{C}},K(k)) = \dim_{\mathbb{R}} V_k.$$

In particular K is the kernel of the map from $\bigoplus_{i=1}^{m} \mathcal{O}(-d_i)$ onto $\mathcal{O}$ determined by the p_i:

$$0 \to K \to \bigoplus_{i=1}^{m} \mathcal{O}(-d_i) \to \mathcal{O} \to 0.$$

By a theorem of Grothendieck [2] the vector bundle K is isomorphic to a direct sum of line bundles (unique up to the order of the factors)

$$K = \bigoplus_{j=1}^{m-1} \mathcal{O}(-a_j).$$

Elementary considerations show that $\sum_{j=1}^{m-1} a_j = \sum_{i=1}^{m} d_i$ and $a_j \geq d_1 \geq 1$. If we could determine the $m - 1$ numbers a_j we would then have

$$\dim_{\mathbb{R}} V_k = \sum_{j=1}^{m-1} (k - a_j + 1)_+$$

and

$$\dim_{\mathbb{R}} D_k = \sum_{j=1}^{m-1} (k - a_j + 1)_+ - \sum_{i=1}^{m} (k - d_i + 1)_+ + k + 1.$$

Note that for $k + 1 \geq \max \{a_j\text{'s}, d_i\text{'s}\}$, we have

$$\dim_{\mathbb{R}} D_k = \sum_{i=1}^{m} d_i - \sum_{j=1}^{m} a_j = 0$$

which provides a desirable bound on the degrees of the solutions (once a_j are known).

In order to give a satisfactory answer, we need only know which sequences of positive integers

$$1 \leq a_1 \leq \cdots \leq a_{m-1}$$

occur. This is answered by the following

<u>Definition</u>: A sequence $1 \leq a_1 \leq \cdots \leq a_{m-1}$ of positive integers is said to be <u>admissible</u> for a given sequence $1 \leq d_1 \leq \cdots \leq d_m$ if

1) $\sum_{j=1}^{m-1} a_j = \sum_{i=1}^{m} d_i$
2) $a_j \geq d_j \quad j = 1,\ldots,m-1$
3) If ℓ is the least index for which

$$a_\ell \neq d_\ell$$

then for $j \geq \ell$ we should have

$$a_j \geq d_{j+1}.$$

Finally, the generic admissible sequence for a given sequence $1 \leq d_1 \leq \cdots \leq d_m$ is that admissible sequence which is largest in the lexicographical ordering--i.e. $(a_1 \ldots a_{m-1}) \leq (a'_1 \ldots a'_{m-1})$ if the first non-zero difference $a'_j - a_j$ is positive.

Theorem 11: Given $p_1, \ldots, p_m$ homogeneous polynomials in $\mathbb{R}[x,y]$ with no common factor above, we have

$$\dim_{\mathbb{R}} V_k = \sum_{j=1}^{m-1} (k - a_j + 1)_+$$

and

$$\dim_{\mathbb{R}} D_k = \sum_{j=1}^{m-1} (k - a_j + 1)_+ - \sum_{i=1}^{m} (k - d_i + 1)_+ + k + 1$$

for some admissible sequence $1 \leq a_1 \leq \cdots \leq a_{m-1}$. Conversely, every admissible sequence arises in this way for some choice of $p_1, \ldots, p_m$ (possible some $p_i \equiv 0$). Moreover in the space $\bigoplus_{i=1}^{m} P_{\mathbb{R}}^{d_i,2} \cong \mathbb{R}^{\sum_{i=1}^{m}(d_i+1)}$ of possible choices of the p_i there is an open dense set (complement has measure zero and is cut out by polynomial equations) of choices having no common factor for which the generic admissible sequence occurs. ◻

The generic admissible sequence gives the smallest possible dimension for the space D given the orders (or degrees) d_i of the partial diferential operators $p_i(\partial)$ (or polynomials p_i).

One can compute many interesting examples. Some of which are of particular relevance in approximation theory and the theory of box splines (see Micchelli and Dahmen [5], [6], and [7] and Chui and Wang [1]). It is to be hoped that the "combinatorial linear algebra" and the examples obtained via the analysis of special configurations of subvarieties currently used in approximation theory to deal with these questions might lead to special and interesting bundles/reflexive sheaves on complex projective spaces.

Bibliography

1. C. K. Chui and R. H. Wang, "Multivariate spline spaces," J. Math. Anal. Appl. (to appear).

2. A. Grothendieck, "Sur la Classification des fibrées holomorphes sur la sphére de Riemann," Amer. J. Math. 79 (1957) 121-138.

3. R. Hartshorne, "On the Classification of Algebraic Space Curves," in Vector Bundles and Differential Equations (Nice 1979), Birkhäuser, Boston, 1980.

4. R. Hartshorne, "Algebraic Geometry," Grad. Texts in Math., Vol. 52, Springer-Verlag, 1977.

5. C. Micchelli and W. Dahmen, "On the local linear independence of translates of a box spline," to appear in Studia Mathematica.

6. C. Micchelli and W. Dahmen, "On the optimal approximation rates for criss-cross finite element spaces," Jour. of Comp. and Applied Math. 10 (1984).

7. C. Micchelli and W. Dahmen, "On the solutions of certain systems of partial difference equations and linear dependence of translates of box splines," IBM Research Report, IBM Thomas J. Watson Research Center, Yorktown Heights, New York.

8. C. Okonek, M. Schneider, and H. Spindler, "Vector bundles on complex projective spaces," Birkhäuser, Boston, 1980.

9. S. Shatz, "On subbundles of vector bundles over $\mathbb{P}^1$," J. of Pure and Appl. Alg. 10 (1977).

10. P. Stiller, "Certain Reflexive Sheaves on $\mathbb{P}^n_{\mathbb{C}}$ and a Problem in Approximation Theory," Trans. of the AMS, Vol. 279, No. 1, Sept. 1983.

Department of Mathematics
Texas A&M University
College Station, Texas 77843

Contemporary Mathematics
Volume 58, Part I, 1986

SMALL POINTS AND TORSION POINTS

Lucien Szpiro

Arakelov intersection theory for compactified divisors on an arithmetic surface allows one to "think geometrically" about number theoretic questions. His idea of putting ad-hoc metric on line bundles has been for example very fruitful in Faltings' proof of the conjectures of Tate, Shafarevich and Mordell.

In particular by analogy with the function field case one sees that to get an "effective Mordell" one needs:

a) the Hodge index theorem

b) small points

I have conjectured these results five years ago. The part a) has been proved and is a trivial consequence of the comparison of Arakelov intersection with Neron-Tate height. (cf §1). This comparison has been recognized by many authors: Arakelov, Faltings, Hriljac, Gross...?

Part b) is still open. I explain it in §2.

The lack of proof for the existence of small points lends naturally to the question: when is the "difference" between two points small? The analysis of this question is done in §3 where we give application to the question of the existence of pairs of distinct points of a curve over $\bar{\mathbb{Q}}$ whose difference is torsion. We show that many curves in the "Kodaira fibration" do not have such a pair of points. We also give a lower bound for the Neron-Tate height of an infinite sequence of points over $\bar{\mathbb{Q}}$ whose Neron-Tate height is bounded. This result applies beautifully to give a simple proof of a theorem of Raynaud about the finiteness of the set of points over $\bar{\mathbb{Q}}$ whose difference from a given point is torsion (under the hypothesis $\omega_{X/C}^2 > 0$ or the stable model has some non irreducible fiber).

I do not give a bibliography, but proof and references to the material here used can be found in my seminar: "Seminaire sur les pinceaux arithmetiques..." in Asterisque spring 1985.

§1. A REVIEW OF ARAKELOV THEORY

1.1 DEGREES AND HEIGHTS

Arakelov theory is based on the following remark: Let L be a projective module of rank one (line bundle) over an order $\underline{\mathcal{O}}$ of a number field, with hermitian positive metrics $|\cdot|_\sigma$ on $L_\sigma = L \underset{\sigma}{\otimes} \mathbb{C}$ for every $\sigma: K \to \mathbb{C}$. Then after picking an element s non zero in L the real number:

$$\deg L = :\log \frac{\# L/s\mathcal{O}}{\prod_\sigma |s|_\sigma} \tag{1}$$

is independent of the choice of s. Now if X_K is a projective variety over K, L_K a line bundle on X_K, for every choice of an integral model (X,L) of (X_K,L_K) over $\underline{\mathcal{O}}_K$ the ring of integers of K, and every choice of a continuous hermitian positive metric on L_σ over X_σ one has a height function $h_L: X_K(\bar{K}) \to \mathbb{R}$ defined by

$$h_L(P) = \frac{1}{[K(P):\mathbb{Q}]} \deg \epsilon_P^* L \tag{11}$$

where $K(P)$ is the field of definition of P and $\epsilon_P : \operatorname{Spec} \underline{\mathcal{O}}_{K(P)} \to X$ the morphism extending $\operatorname{Spec} K(P) \to X_K$ (valuative criterion).

LEMMA 1.1.1 If one changes the integral model of (X_K,L_K), or the continuous metrics on L_σ, the function h_L changes by a bounded function on $X_K(\bar{K})$.

COROLLARY 1.1.1 Let L_K be an ample line bundle on a curve C_K over the number field K. Then for any real number H there exists a point P in $C_K(\bar{K})$ such that $h_L(P) \geqslant H$.

COROLLARY 1.1.2 Let C be a curve over a ring of algebraic integers $\underline{\mathcal{O}}_K$, L a line bundle on C, $|\,|_\sigma$ continuous metrics on L_σ for each σ. Then there exists a real number α and an infinite sequence P_n of points of $C(\bar{K})$ such that $|h_L(P_n)| \leqslant \alpha$.

Lemma 1.1 is a direct consequence of the definitions. Corollary 1.1 and 1.2 come from the fact that there exists an integer n and a morphism $h: C_K \longrightarrow \mathbb{P}^1_k$ such that $h^* \underline{\mathcal{O}}_{\mathbb{P}^1}(1) = L_K^{\otimes n}$. By lemma 1.1 it is enough to have the corollaries on $\mathbb{P}^1$ where one takes the naive height of points:

$(1,N)$ for Corollary 1.1

$(1,\zeta_n)$ $\zeta_n^n = 1$ for Corollary 1.2

1.2 INTERSECTION THEORY ON AN ARITHMETIC SURFACE

Arakelov has developed such a theory which is as close as possible to the geometric case. We give here the most important properties of it.

(I) If P_1 and P_2 are two rational points of the curve (generic fiber) and E_1, E_2 the sections of $C \longrightarrow \operatorname{Spec} \underline{\mathcal{O}}_K$ corresponding, then one has "Green functions" $g_\sigma(Q,R)$ on C_σ for each σ such that

$$(E_1 \cdot E_2) = \log N(E_1 \cap E_2) - \sum_\sigma g_\sigma(P_1,P_2)$$

the functions g_σ are characterised by

a) $\Delta_{\mu_\sigma} g_\sigma(P,Q) = 1 - \delta_P$

where d_{μ_σ} is the metric on C_σ of total volume 1, pulled back of the flat metric on the Jacobian.

b) $\int_{C_\sigma} g_\sigma(P,Q)\, d_{\mu_\sigma}(Q) = 0$

(II) adjunction The choice of the metric in 1)a) enables one to have a canonical metric on the relative dualizing sheaf $\omega_{C/\underline{\mathcal{O}}_K}$ such that for every section E one has:

$$\left(\omega_{C/\mathcal{O}_K} \cdot \underline{\mathcal{O}}_C(E)\right) = -E^2$$

(III) Arakelov intersection is symmetrical

(IV) Comparison to Neron-Tate height

Let denote $\langle\ ,\ \rangle$ the Neron-Tate pairing on the Jacobian. If D_1, D_2 are two Arakelov divisors of degree zero on each component of a fiber then $(D_1 \cdot D_2)_{\text{Arakelov}} = -[K:\mathbb{Q}] \langle x_1, x_2\rangle_{\text{Neron-Tate}}$ where x_i is the point of the Jacobian corresponding to the divisor of degree zero D_i.

COROLLARY 1.2.1 (Hodge Index) The signature of Arakelov pairing is $(+1,-1,-1\ldots-1)$.

The corollary is clear when one knows that Neron-Tate pairing is definite positive and the intersection pairing on fibers is negative semi definite.

COROLLARY 1.2.2 If E is section then

$$(-E^2)\ 2g(2g-2) \leqslant \omega^2_{C/\underline{O}_K}$$

Apply Hodge index to the space generated by $\omega_{C/\underline{O}_K}, E, F$ F = one fiber at infinity. There is a plus sign because $(E+\lambda F)^2 > 0$ if $\lambda \gg 0$.

I saw the next lemma for the first time in G. Faltings' Calculus on arithmetic surfaces. It seems to be classic of capacity theorem:

LEMMA 1.2.1 For any $\epsilon > 0$ there exist an integer $N(\epsilon)$ such that if $P_1 \ldots P_n$, $n \geqslant N(\epsilon)$ are distincts points of C_σ then $\sum_{i \neq j} g_\sigma(P_i, P_j) \leqslant \epsilon n^2$.

This lemma is essential in the proof of the following theorem.

THEOREM (1.2.1) (G.Faltings) Let L be a line bundle on an arithmetic surface $C \to \operatorname{Spec} \underline{O}_K$. Suppose $(L \cdot L) > 0$ and $d^\circ L_K > 0$ then for $n >> 0$ $L^{\otimes n}$ has a positive Arakelov section.

By a positive Arakelov section for a line bundle one means $s \in H^\circ(C, L)$ such that $\int_C \log|s|_\sigma d_{\mu_\sigma} \leqslant 0$ for each σ.

Mixing this theorem with my lemma about sections (Corollary 1.2.2 above) one gets the following:

COROLLARY 1.2.3 Suppose that the genus of C_K is at least two. Then

(I) $\omega_{C/\underline{O}_K}^2 \geqslant 0$

(II) $(\omega_{C/\underline{O}_K} \cdot \underline{O}_X(D)) \geqslant 0$ for any positive Arakelov divisor.

§2. THE CONJECTURE OF SMALL POINTS

Let C be a smooth, geometrically connected curve of genus $g \geqslant 2$ over a number field K. Parshin has remarked that an old construction of Kodaira gives a lot of information about rational points of C. It goes like this:

One builds a smooth non constant fibration $X \xrightarrow{f} C'$ with a commutative diagram

$$\begin{array}{ccc} X & \xrightarrow{\alpha} & C \times C \\ {\scriptstyle f}\downarrow & & \downarrow{\scriptstyle pr.} \\ C' & \xrightarrow{\pi} & C \end{array}$$

such that π is etale, α is finite and ramified only along the diagonal.

Following construction one sees that the fibers X_P, $P \in C'(K)$, of f have good reduction outside $S \cup \{2\}$ where S is the set of bad reduction of C. Moreover there exists a constant A_1 such that:

$$\text{(III)} \qquad \omega^2_{X_P/\underline{O}_K} - d^\circ\alpha(\omega^2_{C/\underline{O}_K} - \tfrac{3}{4}E^2_{\pi(P)})| \leqslant A_1\,[K:\mathbb{Q}]$$

so one sees that to bound $-E^2_{\pi(P)}/[K:\mathbb{Q}]$ (i.e. the "canonical" height of a rational point of C over K - this height is closely related to the Neron-Tate height cf next-) it is enough to bound $\omega^2_{X_P/\underline{O}_K}$.

<u>CONJECTURE</u> Let X be a smooth curve, geometrically connected of genus $g \geqslant 2$ over a number field K and let S be its set of places of bad reduction. Then there exists a point $Q \in X_K(\bar{K})$ such that:

$$-E^2_Q \leqslant (A_2(K) + A_3(S))\ [K(Q):\mathbb{Q}]$$

where A_2 and A_3 are effectively computable.

In the function field case I have proven the existence of point $Q \in X_K(\bar{K})$ such that

$$-E^2_Q \leqslant (2\ \text{genus}(K)-2 + \#S)\ [K(\mathbb{Q}):K]$$

(semi-stable case).

Applying (III) and Corollary 1.2.2. one gets as a corollary of this conjecture a proof of the Mordell conjecture as effective as the constants A_1, A_2, A_3.

To emphasize the importance of the self intersection of the relative dualizing one can keep in mind that one can prove the following statement.

THEOREM 2.1 Let M_g, $g \geqslant 2$, be the moduli space of curves of genus g. Let n be an integer and H a real number. Then the set of points $P \in M_g(K)$ for K a number field of degree at most n, such that the corresponding curve X_P satisfies $\omega_{X_P}^2 \leqslant H$, is finite.

§3 SMALL DIFFERENCE

3.1 THE MUMFORD CONSTANT

Consider the map φ from a curve C of $g \geqslant 2$ to its Jacobian J defined by

$$P \longrightarrow (2g-2)P - \Omega_C^1$$

Let (P_i) be a collection of points of C with images x_i under this map. Applying 1.2 (IV) to the divisors $(2g-2)E_i - \omega_{C/\underline{O}_K} + \emptyset_i$ (where $\emptyset_i$ is a fibral divisor with rational coefficients such that $((2g-2)E_i - \omega_{C/\underline{O}_K} + \emptyset_i \cdot F) = 0$ for any com-component of a fiber of $C \to \operatorname{Spec} \underline{O}_K$) one gets

$$\text{(IV)} \quad \langle X_i \cdot X_j \rangle = \frac{\langle X_i \rangle^2 + \langle X_j \rangle^2}{2g} + \frac{1}{K:\mathbb{Q}} \Big[\omega_{C/\underline{O}_K}^2 \left(\frac{1}{g} - 1\right) - (2g-2)^2 (E_i \cdot E_j) - \frac{\emptyset_i^2 + \emptyset_j^2}{2g} + (\emptyset_i \cdot \emptyset_j)\Big]$$

PROPOSITION 3.1.1. The quantity

$$\frac{1}{K:\mathbb{Q}} \Big[\omega_{C/\underline{O}_K}^2 \left(\frac{1}{g} - 1\right) - (2g-2)^2 (E_P \cdot E_Q) - \frac{\emptyset_P^2 + \emptyset_Q^2}{2g} + (\emptyset_P \cdot \emptyset_Q)\Big]$$

is bounded for $P \neq Q$ P and Q in $C(\bar{K})$. The supremum is called the <u>Mumford constant of C</u> $M(C)$.

This number was introduced by Mumford in 1966, but he did not have a precise formula for it, for lack of Arakelov intersection theory. He showed that for points P_i such that $10\,M(C) \leqslant \langle X_1\rangle^2 \leqslant \langle X_2\rangle^2 \leqslant 2\langle X_1\rangle^2$ the angle for Neron Tate is at least $\frac{\pi}{6}$. The first application of the precise formula above is to show that $M(C)$ is often negative.

PROPOSITION 3.1.2 For any real number a, there exists an infinite number of fibers X_P in the Kodaira fibration such that $M(X_P) < a$.

PROOF: If $M(X_P) \geqslant a$ the formula (IV) implies that there exists a constant $b \in \mathbb{R}$ such that $\omega^2_{X_{P}/\underline{O}_K} \leqslant b[K(P):\mathbb{Q}]$ so by the formula (III) one would get a third constant c such that $-E^2_{\pi(P)} \leqslant c[K(P):\mathbb{Q}]$. This would contradict corollary 1.1.1 if it was true for all but finitely many points $P \in C'(\bar{K})$.

COROLLARY 3.1.1 There exists an infinite number of fibers X_P in the Kodaira fibration such that for any pair of distinct points Q_1 and Q_2 of $X_P(\bar{K})$, Q_1-Q_2 is not a torsion point in the Jacobian of X_P.

PROOF: If Q_1-Q_2 is torsion then $x_1 = x_2$ and hence $M(X_P) \geqslant 0$ by formula (IV).

3.2 INFINITELY MANY POINTS OF THE "SAME SIZE".

3.2.1 The following theorem, together with Corollary 1.1.2 answers a question of Bogomulov.

THEOREM 3.2.1 Let a be a real number and X_K a smooth geometrically connected curve of genus $g \geqslant 2$ over a number field K. Suppose that X_K has semi-stable reduction and that one has an infinite sequence of distinct points (P_i) in $X_K(\bar{K})$ whose images x_i by the morphism φ of 3-1, have Neron Tate height smaller than a. Then $\omega^2_{X/\underline{O}_K} \leqslant a\left(\frac{g+1}{g-1}\right) + C(S)$ where $C(S)$ depends only on the intersection matrix on the bad fibers S, of X over $\underline{O}_K$. (In particular $C(S) = 0$ if $S = \emptyset$.)

Looking at the formula (IV) it is enough to prove (by adding $\frac{n^2-n}{2}$ terms like it)

a) $\forall \varepsilon > 0 \quad -\sum_{i\neq j \leqslant n} (E_i \cdot E_j) \leqslant \varepsilon [K':\mathbb{Q}] n^2$ for $n >> 0$ and K' a field of rationality of $P_1 \cdot\cdot P_n$

b) $\sum_{i\neq j \leqslant n} (-\frac{\emptyset_i^2 + \emptyset_j^2}{2} + (\emptyset_i \cdot \emptyset_j) \leqslant [K':\mathbb{Q}]\, C(S) n^2$

- For a) one sees that $-(E_i \cdot E_j) \leqslant \sum_\sigma g_\sigma(P_i, P_j)$ so one concludes by the lemma 1.2.1 (capacity theory).

- For b) I have proved in loc·cit in the introduction that there exists $C(S)$ such that $-\frac{\emptyset_i^2 + \emptyset_j^2}{2} + (\emptyset_i \cdot \emptyset_j) \leqslant [K':\mathbb{Q}]\, C(S)$.

One should notice that if an infinite subsequence of the (P_i) all pass by the same component of bad fibers than $\emptyset_1 = \emptyset_2 = \cdots = \emptyset_n = \cdots$ so $-\frac{\emptyset_i^2 + \emptyset_j^2}{2g} + (\emptyset_i \cdot \emptyset_j) \leqslant \frac{\emptyset_i^2 + \emptyset_j^2}{2g} (1 - \frac{1}{g}) \leqslant 0$. A variation on this theme will be very useful for the next theorem (when a = 0)

<u>THEOREM 3.2.2</u> Let X_K be a semi-stable curve like in theorem 3.2.1. Suppose $\omega^2_{X/\underline{O}_K} > 0$. Then there is only a finite number of points in $X_K(\bar{K})$ whose difference with a given point is torsion in the Jacobian.

This theorem without the hypothesis $\omega^2_{X/\underline{O}_K} > 0$ has been proved, by quite a different method, by M. Raynaud. In fact one can see by analyizing the proof below that the only cases of the theorem of Raynaud that I do not get are: the bad (semi-stable) fibers have only one component which is not a rational curve of self intersection -2 and the possibly infinite sequence of points (P_i) such that their difference is torsion also satisfied:

a) all the P_i pass by this non rational component

b) $\forall_i, \exists n_i \quad \underline{O}_{X_K}((2g-2)n_i P_i) = \Omega^{1\,\otimes n_i}_{X_{\bar{K}}/\bar{K}}$

c) $\omega^2_{X_{K/\underline{O}_K}} = -E_i^2 = 0$

By Theorem 2.1 one sees that $\omega^2_{X/\underline{O}_K} = 0$ can be satisfied only for a finite number of curves of a given genus order H. In fact I would conjecture that $\omega^2_{X/\underline{O}_K} > 0$ if X has some bad reduction. In the geometric case $\omega^2_{X/\underline{O}_K} = 0$ implies that the family is constant. There is no such notion for an arithmetic pencil. I would be surprised if there exists smooth curves over $\underline{O}_K$ with $\omega^2_{X/\underline{O}_K} = 0$. If they exist they deserve the name of "constant" curves. To be more philosophical, and because it is the case for function fields over finite fields, I expect a relation between $\omega^2_{X/\underline{O}_K}$ and the focus of ordinary reduction.

PROOF OF THEOREM 3.2 We start like in the proof of 3.2.1, applying the formula (IV). The terms in $-(E_i \cdot E_j)$ are bound by ϵn^2, but we have to do better than in 3.1 for the term in $\emptyset_i, \emptyset_j$. If m is big enough and $E_1 - E_0$ is of m torsion exactly (E_0 rational over K, E_1 is rational over a bigger field) the all $\mathrm{Gal}(\bar{K}/K)$ orbit of E_1 also gives points whose difference E_0 is m torsion. This orbit makes a $\mathbb{Z}/m\mathbb{Z}$ module of the form $(\mathbb{Z}/m\mathbb{Z})^r$ $r \leqslant 2g$. For every place $\nu \in S$ of bad reduction, there will be an integer r_ν, $0 < r_\nu \leqslant r$ such that m^{r_ν} of the conjugates of $E_1 - E_0$ will be in each possible component of the special fiber of the Neron-Model.

Hence $(\emptyset_i - \emptyset_j)^2$ will be zero if $E_i - E_0$ and $E_j - E_0$ lie in the same component of the Neron-Model in ν. So like

$$\sum_{i \neq j \leqslant m^r} - \frac{\emptyset_i^2 + \emptyset_j^2}{2g} + (\emptyset_i \cdot \emptyset_j) = \sum_{i \neq j \leqslant m^r} -\frac{1}{2}(\emptyset_i - \emptyset_j)^2 + \frac{\emptyset_i^2 + \emptyset_j^2}{2}\left(1 - \frac{1}{g}\right)$$

is bounded by

$$\sum_{\nu t S} - \frac{1}{2} \sum_{i \neq j \leqslant m^r} (\emptyset_i - \emptyset_j)^2$$

We have a constant α such that

$$\left(\Sigma - \frac{\emptyset_i^2 + \emptyset_j^2}{2g} + (\emptyset_i \cdot \emptyset_j)_\nu\right) \leqslant \alpha\left(\sum_{i=1}^{m^{r-r_\nu}} \sum_{j=1}^{i} j\right) m^r \quad N(\nu)$$

$$\leqslant \alpha m^r \ N(\nu) \quad m^{2(r-r_\nu)} + 1$$

If ν_0 is a place of K in S one has $[K:\mathbb{Q}] \geqslant m^r \sum_{\nu|\nu_0} N(\nu)$. One gets that, putting $n = m^r$, $\sum_{i\neq j<n} -\frac{\emptyset_i^2 + \emptyset_j^2}{2g} + (\emptyset_i \cdot \emptyset_j) \leqslant \varepsilon[K':\mathbb{Q}]n^2$ so applying formula (IV) $(n^2-n)/_2$ time one gets

$\omega_{X/\underline{O}_K}$ $(1-\frac{1}{g})\ |x|^2\ (\frac{1}{g} - 1)\ + \varepsilon$ for each $\varepsilon > 0$ where $|x|^2$ is the Neron Tate height of all the P_i (corresponding points x_i are all equal). So if $|x|^2 \neq 0$ one gets a contradiction, and if $|x|^2 = 0$ one gets $\omega^2_{X/\underline{O}_K} = 0$.

Ecole normale superieure
Centre de Mathematiques ERA 589
45 rue d'Ulm 75005 Paris France

Columbia University
Department of Mathematics
New York, NY 10027 USA

Contemporary Mathematics
Volume 58, Part I, 1986

GREEN'S THEORY OF CHERN CLASSES AND THE RIEMANN - ROCH FORMULA

DOMINGO TOLEDO AND[1]
YUE LIN L. TONG[1]

ABSTRACT. We explain a Theory of Chern classes in de Rham cohomology for coherent sheaves on complex manifolds developed in the unpublished thesis of H.I. Green. We then prove the Hirzebruch-Riemann-Roch formula for these classes, thus giving an affirmative answer to a question posed by Atiyah, Hirzebruch and Singer.

The main purpose of this paper is to publicize the unpublished thesis of H.I. Green on Chern classes for coherent sheaves on complex manifolds [5]. We do this in two ways. First, we give an exposition of the thesis. Then we show it can be combined with a natural extenssion of our proof of the Hirzebruch-Riemann-Roch formula (HRR) for vector bundles [12] to prove HRR for coherent sheaves, and Chern classes in de Rham cohomology. Thus we settle a long-standing question, possed implicitly by Atiyah and Hirzebruch in [3] and explicitly by Singer in [11].

A major difficulty in the two subjects considered here is that they cannot be deduced from the known results on vector bundles because, in contrast with the algebraic category, the question of existence of global resolutions of a coherent sheaf by locally free sheaves is wide open. We refer to a recent paper by Schuster [10] for the present status of this problem.

It is then natural to consider substitutes for global resolutions by locally free holomorphic sheaves. This can be done in at least two directions. One is by working in the C^∞ or real analytic category, and global resolutions by vector bundles then exist, as prove by Atiyah and Hirzebruch in [2]. This gives an immediate definition of Chern classes, but, since the holomorphic structure is lost, it seems quite difficult to prove HRR in this way.

Another direction is to stay in the complex analytic category, but use local objects with a more complicated global structure than vector bundles. We proposed

---

[1] Partially supported by the National Science Foundation.

such an object, the twisting cochain, in [13], and were able to define Atiyah-Chern classes [7] and prove HRR for these classes [8]. We remark that twisting cochains serve two different purposes. One is to resolve the sheaf of the diagonal in order to follow Lefschetz's original idea [6], and for this purpose seems to be optimal- cf. §4. The other is to resolve the coefficient sheaf, and for this purpose other substitutes are desirable.

Green develops another substitute in this second direction, one that is particularly useful from the point of view of extending additive functions from vector bundles to coherent sheaves. Its structure is close enough to vector bundles that the formulas for Chern classes presented by Bott at the 1971 ELAM [4] can be applied to define Chern classes in de Rham cohomolgy, and to prove that they agree with those of Atiyah-Hirzebruch. Our proof of HRR [12] can also be applied in this context to prove our main theorem in §4.

Green's ideas are intimately connected with K-theory, in a way that makes us believe that, with proper care and understanding, can be applied to answer any possible Riemann-Roch question. In particular combined with the basic ideas of [9] just as the present paper is combined with [12], they should settle the Grothendieck-Riemann-Roch problem in any cohomology theory.

Since Green's work is unpublished, we devote most of the paper to its exposition. We present only those aspects that are relevant for our proof of HRR, and modify his presentation to bring out clearly the structure of his resolution that we find most relevant for applications. In §1 we state, without proof, his main theorem. In 2 we explain the Chern classes. In §3 we give, except for some computational details, the proof of the theorem in §1. The last section, §4, is the only non-expository one, where we prove HRR. We remark that it could be proved in another way (comparing definitions of Chern classes) that we had already proved the theorem of §4 in [8], but we had no way of proving this without Green's thesis, and the approach in §4 is much simpler.

We are very grateful to H.I. Green for making his work available to us, and to N.R. O ' Brian for many valuable comments. We also thank the audience at Lefschetz Centennial Conference for their reaction; particularly J. Eells for his encouragement to write this paper in its present form, and O. Forster and P. Griffiths for informing us of the current status of the question of existence of global resolutions.

§1. SIMPLICIAL VECTOR BUNDLES.

In order to explain Green's construction we introduce a notion of "simplicial vector bundle" over the nerve of a covering. This term has been used in the literature in several ways, all different from the one here.

The context is the following. We suppose we are given a manifold M (for our purpose mainly a complex manifold with structure sheaf $\mathcal{O}_M$, but could also be real analytic or C^∞ manifold with $\mathcal{O}_M$ = sheaf of real analytic, respectively C^∞ functions), and an open covering $\mathfrak{U} = \{U_\alpha\}$. We write $N\mathfrak{U}$ for the nerve of $\mathfrak{U}$. It $\sigma = \langle\alpha_0 \ldots \alpha_p\rangle$ is a simplex of $N\mathfrak{U}$, we write U_σ for $U_{\alpha_0} \cap \ldots \cap U_{\alpha_p}$.

We assume given the following data:

(SB 1) For each simplex σ of $N\mathfrak{U}$, a free $\mathcal{O}_{U_\sigma}$ - module E_σ with basis $e_\sigma = \{e_\sigma^1, \ldots, e_\sigma^{k(\sigma)}\}$.

(SB 2) For each simplex σ and each pair of vertices α,β of σ, an isomorphism ${}^\sigma a_{\alpha\beta} : E_\sigma \to E_\sigma$ linear over $\mathcal{O}_{U_\sigma}$.

(SB 3) The isomorphisms ${}^\sigma a_{\alpha\beta}$ satisfy the cocycle condition in the sense that for any simplex σ and any three vertices α,β,γ of σ, the identify

$$ {}^\sigma a_{\alpha\beta} \, {}^\sigma a_{\beta\gamma} = {}^\sigma a_{\alpha\gamma} $$

holds.

(SB 4) Whenever $\sigma < \tau$ there is an inclusion of the basis e_σ in e_τ. Thus if we let $e_{\sigma,\tau}$ denote the complement of e_σ in e_τ, and $E_{\sigma,\tau}$ denote the free $\mathcal{O}_{U_\tau}$ - module with basis $e_{\sigma,\tau}$, we have a short exact sequence

$$ 0 \to E_\sigma|_{U_\tau} \to E \to E_{\sigma,\tau} \to 0. $$

(SB 5) For any pair of simplices $\sigma < \tau$ and any two vertices α,β of σ, the matrix ${}^\tau a_{\alpha\beta}$ is of the form

$$ {}^\tau a_{\alpha\beta} = \begin{bmatrix} {}^\sigma a_{\alpha\beta} & * \\ 0 & 1 \end{bmatrix}. $$

We call such a collection $\{E_\sigma, {}^\sigma a_{\alpha\beta}\}$ a *one-cocycle of compatible isomorphisms* and denote it simply by (E,a). We can also think of it as a compatible collection of one-cocycles of isomorphisms, i.e., a one-cocycle ${}^\sigma a_{\alpha\beta}$ over each U_σ, all of these being compatible in the sense of (SB 5). More looseley, we could think of it as a compatible collection of vector bundles. But this can only be made precise by specifying condition (SB 5) on the transition functions.

By a *simplicial vector bundle* we mean simply a one-cocycle of compatible isomorphisms. (Strictly speaking modulo an obvious equivalence relation, which we will not use explicitly).

Note that a vector bundle is a special case of a simplicial vector bundle.

Namely, given a one-cocycle $\{a_{\alpha\beta}\}$ in the usual sense, let ${}^{\sigma}a_{\alpha\beta} = a_{\alpha\beta}|U_{\sigma}$.

By a complex of simplicial vector bundles we mean a sequence $E = \{E^{-k},\ldots,E^{-1},E^{o}\}$ and differentials $b^{l}: E^{l} \to E^{l+1}$ which are bundle mappings in the sense tnat $b^{l} = \{{}^{\sigma}b^{l}_{\alpha}\}$, where ${}^{\sigma}b^{l}_{\alpha}: E^{l}_{\sigma} \to E^{l+1}_{\sigma}$ and ${}^{\sigma}b^{l}_{\alpha}\ {}^{\sigma}a^{l}_{\alpha\beta} = {}^{\sigma}a^{l+1}_{\alpha\beta}\ {}^{\sigma}b^{l}_{\beta}$.

We say that a complex of simplicial vector bundles *resolves* a coherent sheaf $\mathcal{S}$ on M if for each σ and $\alpha \in \sigma$ there is exact sequence of sheaves

$$0 \to E^{-k}_{\sigma} \xrightarrow{{}^{\sigma}b_{\alpha}} E^{-k+1}_{\sigma} \to \ldots \xrightarrow{{}^{\sigma}b_{\alpha}} E_{\sigma} \to \mathcal{S}|_{U_{\sigma}} \to 0 .$$

(to simplify later notations, we will always assume that the length k of the resolutions is always $n = \dim_{\mathbb{C}} M$).

Green's Theorem 1: Every coherent sheaf $\mathcal{S}$ over M has a resolution by a complex of simplicial vector bundles. Moreover the resolution can be chosen so that for each $\sigma < \tau$ the complex $E_{\sigma,\tau}$ as in (SB 4) is a direct sum of elementary sequences.

Recall that an elementary sequence is a complex of the form $0 \to \ldots 0 \to E \xrightarrow{id} E \to 0 \to \ldots$. Thus $E_{\sigma}|_{U_{\tau}}$ and E_{τ} have the same simple homotopy type.

§2. CHERN CLASSES.

The relevance of simplicial vector bundles is that they allow the extension of certain additive natural transformations from vector bundles to coherent sheaves. We illustrate this extension first for the simple example of the determinant line bundle (or first Chern class in Pic M), then for Chern classes in de Rham cchomology. The formulation of the most general theorem of this nature is left to the reader.

Let (E,a) be a simplicial vector bundle. We claim that the cochain

$$\left\{\det\left[{}^{\langle\alpha\beta\rangle}a_{\alpha\beta}\right]\right\} \in C'(\mathfrak{U},\mathcal{O}^*)$$

is a cocycle. Namely, for any simplex $\langle\alpha\beta\gamma\rangle$ we have by SB 5 the identities

$$\det\left[{}^{\langle\alpha\beta\rangle}a_{\alpha\beta}\right]|U_{\langle\alpha\beta\gamma\rangle} = \det\left[{}^{\langle\alpha\beta\gamma\rangle}a_{\alpha\beta}\right]$$

$$\det\left[{}^{\langle\alpha\gamma\rangle}a_{\alpha\gamma}\right]|U_{\langle\alpha\beta\gamma\rangle} = \det\left[{}^{\langle\alpha\beta\gamma\rangle}a_{\alpha\gamma}\right]$$

$$\det\left[{}^{\langle\beta\gamma\rangle}a_{\beta\gamma}\right]|U_{\langle\alpha\beta\gamma\rangle} = \det\left[{}^{\langle\alpha\beta\gamma\rangle}a_{\beta\gamma}\right]$$

and by SB 3 we also have

$$\det\left[{}^{<\alpha\beta\gamma>} a_{\alpha\beta}\right] \det\left[{}^{<\alpha\beta\gamma>} a_{\beta\gamma}\right] = \det\left[{}^{<\alpha\beta\gamma>} a_{\alpha\gamma}\right].$$

Hence over $U_{<\alpha\beta\gamma>}$ we get the cocycle identity

$$\det\left[{}^{<\alpha\beta>} a_{\alpha\beta}\right] \det\left[{}^{<\beta\gamma>} a_{\beta\gamma}\right] = \det\left[{}^{<\alpha\gamma>} a_{\alpha\gamma}\right]$$

as claimed. Thus a simplicial vector bundle (E,a) has a determinant bundle $\det(E,a)$. If $\mathfrak{S}$ is a coherent sheaf and $(E^{\cdot},a)$ is a simplicial resolution of $\mathfrak{S}$, by standard arguments (using, e.g., [7], Prop 1.10 and §3), the line bundle

$$\det(\mathfrak{S}) = \prod_{l} \det(E^{l},a)^{(-1)^{l}}$$

is will defined (independent of $(E^{\cdot},a)$).

The first Chern class of $\mathfrak{S}$ in de Rhamm cohomology is simply $c_1(\det \mathfrak{S})$ and is explicitly represented by the cocycle

$$\Sigma(-1)^{l} \text{ d log det } {}^{<\alpha\beta>} a^{l}_{\alpha\beta} \in C^{1}(\mathfrak{U}, \Omega^{1}_{\text{closed}}).$$

Direct computation from Green's resolution shows that this is the same cocycle given by the general formulas of [7]. This cocycle has also appeared, from different points of view, in unpublished work of other authors. But Green's is the first construction of Cech - de Rham cocycles for higher Chern classes.

Now to the general construction. This is based on the formal properties of the Cech - de Rham cocycles introduced by Bott in [4] to represent the Chern classes of (holomorphic) vector bundles in $H^{*}(M,\mathbb{C})$. Namely, we follow Bott's recipe applied to the followings situation. Start with a holomorphic vector bundle E over M, trivialized over each $U_{\alpha} \in \mathfrak{U}$ and with transition functions

$$a = \{a_{\alpha\beta}\},\ a_{\alpha\beta}\colon U_{<\alpha\beta>} \to GL(m,\mathbb{C}),$$

and a $GL(m,\mathbb{C})$-invariant polynomial φ of degree p on $M \times M$ complex matrices. Give $E|U_{\alpha}$ the flat connection corresponding to the trivialization, i.e., ∇_{α} (section) = d (corresponding vector function). Interpolate the connections and fibre integrate as described in [4]. The result is easily checked to give the following structure:

For each invariant polynomial φ of degree p a sequence of Cech - de Rham cochains

$$\omega^{k}_{\varphi}(a) \in C^{k}(\mathfrak{U}, \Omega^{2p-k}),\ k = 1,\dots,p,$$

with the properties:

(B 1) For each $x \in U_{<\alpha_0 \dots \alpha_k>}$, $\omega^{k}_{\varphi}(a)_{\alpha_0 \dots \alpha_k}(x)$ = universal polynomial expresion in

$$a_{\alpha_{i-1}\alpha_i}(x),\ a^{-1}_{\alpha_{i-1}\alpha_i}(x),\ d\, a_{\alpha_{i-1}\alpha_i}(x),$$

depending only on φ and k. Thus we write $\omega^k_\varphi(a_{\alpha_0\alpha_1},\dots,a_{\alpha_{k-1},\alpha_k})$ for $\omega^k_\varphi(a)_{\alpha_0\dots\alpha_k}$ to emphasize that this expresion is defined whenever we are given an open set U in a complex manifold and holomorphic functions $a_{\alpha_i\alpha_j}: U \to GL(m,\mathbb{C})$. Moreover, ω^k_φ is natural under local biholomorphic maps $V \to U$.

(B 2) The sequence ω^k_φ is a universal Cech - de Rham cocycle in the sense that whenever we are given $a_{\alpha_i\alpha_j}: U \to GL(m,\mathbb{C})$ as in (B 1) which satisfy the cocycle condition $a_{\alpha_i\alpha_j}\, a_{\alpha_j\alpha_k} = a_{\alpha_i\alpha_k}$, the identity

$$\begin{aligned}&\omega^k(a_{\alpha_1\alpha_2},\dots, a_{\alpha_k\alpha_{k+1}})\\ &+\Sigma(-1)^i\,\omega^k_\varphi(a_{\alpha_0\alpha_1},\dots, a_{\alpha_{i-1}\alpha_{i+1}},\dots, a_{\alpha_k\alpha_{k+1}})\\ &+(-1)^{k+1}\,\omega^k_\varphi(a_{\alpha_0\alpha_1},\dots, a_{\alpha_{k-1}\alpha_k})\\ &= d\,\omega^{k+1}_\varphi(a_{\alpha_0\alpha_1},\dots, a_{\alpha_k\alpha_{k+1}}) .\end{aligned}$$

Thus, universally $\delta\omega^k = d\,\omega^{k+1}$.

(B 3) $$\omega^p_\varphi(a_{\alpha_0\alpha_1},\dots, a_{\alpha_{p-1}\alpha_p}) = \varphi(a^{-1}_{\alpha_0\alpha_p}\, d\,a_{\alpha_0\alpha_1}\wedge\dots\wedge d\,a_{\alpha_{p-1}\alpha_p}) ,$$

the universal representative for the Atiyah - Chern class of E [1], cf [12].

(B 4) Stability:

$$\omega^k_\varphi(a_{\alpha_0\alpha_1},\dots, a_{\alpha_{k-1}\alpha_k}) = \omega^k_\varphi\left(\begin{pmatrix} a_{\alpha_0\alpha_i} & * \\ 0 & 1\end{pmatrix},\dots, \begin{pmatrix} a_{\alpha_{k-1}\alpha_k} & * \\ 0 & 1\end{pmatrix}\right)$$

(here we think of φ as defined on matrices of any size in the standard way).

It is proved in [4] that the cocycle $\omega'_\varphi(a) + \dots + \omega^p_\varphi(a)$ in the total complex of $C^*(\mathfrak{U},\Omega^*)$ represents the Chern class of E corresponding to φ in $H^{2p}(M,\mathbb{C})$, and its "leading term in the Hodge filtration", the δ-cocycle $\omega^p_\varphi(a)$, referents the Atiyah - Chern class in $H^p(M,\Omega^p)$.

The beauty of this construction is that it extends immediately to simplicial vector bundles. Let (E,a) be a simplicial vector bundle (and to avoid certain technicalities, assume the index set of $\mathfrak{U}$ has a partial order that restricts to a total order in each simplex, and consider only increasing simplices). Define (cf B 1)

$$\omega^k_\varphi(a)_{\alpha_0\dots\alpha_k} = \omega^k_\varphi({}^\sigma a_{\alpha_0\alpha_1},\dots, {}^\sigma a_{\alpha_{k-1}\alpha_k}) ,$$

where $\sigma = \langle\alpha_0\dots\alpha_k\rangle$.

Just as with the discussion of the determinant bundle, we have:

THEOREM. *The* $\omega_{\varphi}^{k}(a)$, $k=1,\ldots,p$, *form a Chech - de Rham cocycle.*

Proof. We have to check

$$(*) \quad (\delta\omega^{k}(a))_{\alpha_0\ldots\alpha_{k+1}} = d\,\omega_{\varphi}^{k+1}(a)_{\alpha_0\ldots\alpha_{k+1}} .$$

Let $\tau = \langle\alpha_0\ldots\alpha_{k+1}\rangle$ and $\sigma_i = \langle\alpha_0\ldots\hat{\alpha}_i\ldots\alpha_{k+1}\rangle$.

By SB 5 and B 4 we get

$$\omega_{\varphi}^{k}({}^{\sigma_i}a_{\alpha_0\alpha_1},\ldots,{}^{\sigma_i}a_{\alpha_{i-1}\alpha_{i+1}},\ldots,{}^{\sigma_i}a_{\alpha_k\alpha_{k+1}})\,|\,U_\tau$$

$$= \omega_{\varphi}^{k}({}^{\tau}a_{\alpha_0\alpha_1},\ldots,{}^{\tau}a_{\alpha_{i-1}\alpha_{i+1}},\ldots,{}^{\tau}a_{\alpha_k\alpha_{k+1}}) ,$$

(obvious modification for $i = 0,k+1$), and from SB 3 and B 2 :

$$(\delta\omega_{\varphi}^{i}({}^{\tau}a))_{\alpha_0\ldots\alpha_{k+1}} = d\,\omega_{\varphi}^{k}({}^{\tau}a)_{\alpha_0\ldots\alpha_{k+1}} .$$

Combinining these two identities with the definition, (*) follows, and the theorem is proved.

We can now define the Chern classes of a simplicial vector bundle (E,a) to be the cohomology classes of the above cocycles. Thus the p'th Chern class c_p is obtained by taking φ = trace of p'th exterior power , the p'th component of the Chern character by taking the suitable multiple of φ = trace of p'th power, etc. The total Chern class (resp. Chern character) of a coherent sheaf $\mathfrak{F}$ is then defined to be the alternating product (resp. alternating sum) of those of the simplicial bundles in a simplicial resolution $(E^{\cdot},a)$. Standard arguments show that the resulting cohomology class is independent of the resolution. Moreover, if we tensor $(E^{\cdot},a)$ with the sheaf $\mathcal{O}_{an}$ of real analytic functions on M we obtain a simplicial resolution of $\mathfrak{F}\otimes\mathcal{O}_{an}$ (flatness of $\mathcal{O}_{an}$ over $\mathcal{O}$). Thus comparing this with a global resolution of $\mathfrak{F}\otimes\mathcal{O}_{an}$ as in Atiyah-Hirzebruch [2,3] we obtain that these classes agree with those defined via global real-analytic resolutions.

We write

$c_p^{G}(\mathfrak{F})$ for the (cohomology class of the) Chech - de Rham cocycle just defined,

$c_p^{A}(\mathfrak{F})$ for (the cohomology class of) its component of bidegree $p-p$,

$c_p^{A-H}(\mathfrak{F})$ for the class of Atiyah-Hirzebruch.

The discussion of Chern classes can be sumarized as follows:

Green's Theorem 2: The classes $c_p^{G}(\mathfrak{F}) \in H^{2p}(M,\mathbb{C})$ have the properties:

1) $c_p(\mathfrak{F})$ is represented by a total cocycle in $\oplus\, C^{k}(\mathfrak{U},\Omega^{2p-k})$, $1 \leq k \leq p$,

whose component of bidegree $p-p$ gives $c_p^A(\mathcal{S}) \in H^p(M,\Omega^p)$. Thus the $c_p^G(\mathcal{S})$ have the same Hodge filtration as the Chern classes of holomorphic vector bundles.

2) $c_p^G(\mathcal{S}) = c_p^{A-H}(\mathcal{S})$ in $H^{2p}(M,\mathbb{C})$.

We denote the common value of c_p^G and c_p^{A-H} simply by $c_p(\mathcal{S})$.

A similar discussion holds for the Chern character $ch(\mathcal{S})$.

§3. CONSTRUCTION OF THE RESOLUTION.

We now present Green's construction. He starts with a twisting cochain for $\mathcal{S}$ as in [13] and successively modifies it to obtain a complex of simplicial vector bundles. We recall that a twisting cochain for $\mathcal{S}$ means

1) over each $U_\alpha \in \mathfrak{U}$ a free resolution

$$0 \to E_\alpha^{-n} \xrightarrow{a_\alpha^{0,1}} \cdots \xrightarrow{a_\alpha^{0,1}} E^0 \to \mathcal{S}|U_\alpha \to 0,$$

2) over each U_σ, $\sigma = \langle\alpha_0\ldots\alpha_k\rangle$, a bundle map $a_{\alpha_0\ldots\alpha_k}^{k,1-k} : E_{\alpha_k}^{\cdot}|U_\sigma \to E_{\alpha_0}^{\cdot}|U_\sigma$ of degree $1-k$,

3) the cochain $a = a^{0,1} + a^{1,0} + a^{2,-1} + \ldots$ satisfies the identities $\delta a + a \cdot a = 0$, $a_{\alpha\alpha}^{1,0} = 1$.

We refer to [7],§1 for details of the meaning of $\delta a + a \cdot a = 0$, and for the simple theorem that a coherent sheaf can be resolved by a twisting cochain.

To be able to go from a twisting cochain to a complex of simplicial vector bundles we need a concept that interpolates between them, namely a *simplicial twisting cochain* on $N\mathfrak{U}$. This means:

(STC 1) For each vertex α and each $\sigma \geqslant \langle\alpha\rangle$ a free resolution: $E_{\sigma,\alpha}^{\cdot}$ of $\mathcal{S}|U_\sigma$: $0 \to E_{\sigma,\alpha}^{-n} \xrightarrow{{}^\sigma a_\alpha^{0,1}} \cdots \xrightarrow{{}^\sigma a_\alpha^{0,1}} E_{\sigma,\alpha}^0 \to \mathcal{S}|U_\sigma \to 0$

(STC 2) For each $\sigma = \langle\alpha_0\ldots\alpha_p\rangle$ a twisting cochain ${}^\sigma a = {}^\sigma a^{0,1} + {}^\sigma a^{1,0} + a^{2,-1} + \ldots$ acting on the resolutions $E_{\sigma,\alpha_i}^{\cdot}$.

(STC 3) Whenever $\sigma < \tau$ and α a vertex of σ, a short exact sequence of free modules $0 \to E_{\sigma,\alpha}^{\cdot}|U_\tau \to E_{\tau,\alpha}^{\cdot} \to E_{\sigma,\tau,\alpha}^{\cdot} \to 0$ as in (SB 4), §1.

(STC 4) Whenever $\sigma < \tau$ and $\alpha_0,\ldots,\alpha_k$ any vertices of σ (repetitions allowed!), the matrix ${}^\tau a_{\alpha_0\ldots\alpha_k}^{k,1-k}$ is of the form:

i) for $k = 0$, ${}^\tau a_\alpha^{0,1} = {}^\sigma a_\alpha^{0,1} \oplus$ elementary sequences,

ii) for $k = 1$ $\quad {}^\tau a_{\alpha\beta}^{1,0} = \begin{bmatrix} {}^\sigma a_{\alpha\beta}^{1,0} & * \\ 0 & 1 \end{bmatrix}$

iii) for $k > 1$, $\tau a^{k,1-k}_{\alpha_0\ldots\alpha_k} = \begin{bmatrix} {}^{\sigma}a^{k,1-k}_{\alpha_0\ldots\alpha_k} & * \\ 0 & 0 \end{bmatrix}$.

Simplicial twisting cochains include both twisting cochains and complexes of simplicial vector bundles. Namely, a twisting cochain gives an STC by taking $E^{\cdot}_{\sigma,\alpha} = E^{\cdot}_{\alpha}|U_{\sigma}$ and ${}^{\sigma}a = a|U_{\sigma}$. A complex of simplicial vector bundles is an STC such that

1) $E^{\cdot}_{\sigma,\alpha}$ is independent of α, denoted simply by $E^{\cdot}_{\sigma}$,

2) ${}^{\sigma}a^{k,1-k} = 0$ for $k > 1$.

We can now present the inductive construction. It will be convenient to temporarily drop the assumption that the $E^{\cdot}_{\sigma,\alpha}$ in an STC are negatively graded. Thus we allow

$$\ldots 0 \to 0 \to E^{-n+r}_{\sigma,\alpha} \to \ldots \to E^{o}_{\sigma,\alpha} \to \ldots E^{r}_{\sigma,\alpha} \to 0 \to 0 \to \ldots$$

We assume given an STC $(E^{o}_{\sigma,\alpha}, {}^{\sigma}a)$ that also satisfies:

*1) For $j \geqslant 1$, the modules $E^{j}_{\sigma,\alpha}$ are independent of α,

*2) For $j \geqslant 1$, ${}^{\sigma}a^{q,1-q}|E^{j}_{\sigma} = 0$ for $q > 1$.

To such an STC we apply a modification denoted $(M(E)^{\cdot}_{\sigma,\alpha}, {}^{\sigma}M(a))$ defined as follows. Fix a simplex $\sigma = <\alpha_o\ldots\alpha_p>$ and to simplify the notation drop the suffix α (i.e., write $\sigma = <o\ldots p>$, $E^{\cdot}_{\sigma,i}$ for $E^{\cdot}_{\sigma,\alpha_i}$, etc). Let $M(E)$ be defined by

$$M(E)^{j}_{\sigma,i} = E^{j-1}_{\sigma,i} \quad \text{for} \quad j < 0 \quad \text{or} \quad j > 1 \; ,$$

$$M(E)^{j}_{\sigma,i} = E^{o}_{\sigma,o} \oplus E^{o}_{\sigma,1} \oplus\ldots\oplus E^{j-1}_{\sigma,i} \oplus\ldots\oplus E^{o}_{\sigma,p} \quad \text{for} \quad j = 0,1 \; .$$

The new twisting cochain $M(a)$ is defined by $M(a) = a$ in dimensions $j \neq 0,1$. In dimension 0:

$${}^{\sigma}M(a)^{o,1}_{i} = \begin{Bmatrix} 1 & & & & \\ & \ddots & & & \\ & & 1 & {}^{\sigma}a^{o,1}_{i} & \\ & & & \ddots & \\ & & & & 1 \end{Bmatrix} ,$$

${}^{\sigma}M(a)^{1,o}_{ji}$ has $r-s$ entry (r = row, s = column) , given by

$$\begin{cases} \delta_{rs} & r \neq i,j \\ {}^{\sigma}a_i^{o,1} & r = i,\ s = i \\ -{}^{\sigma}a_{is}^{1,o} & r = i,\ s \neq i \\ {}^{\sigma}a_{ji}^{1,o} & r = j,\ s = i \\ {}^{\sigma}a_{jis}^{2,-1} & r = j,\ s \neq i \end{cases}$$

if $i \neq j$, and ${}^{\sigma}M(a)_{ii}^{1,o} = 1$,

${}^{\sigma}M^{q,1-q}_{j_o \cdots j_q}$, for $q \geqslant 2$, is given by the row vector

$$\left[(-1)^{q+1}\, {}^{\sigma}a^{q+1,-q}_{j_o,\ldots,j_q,o},\ (-1)^{q+1}\, {}^{\sigma}a^{q+1,-q}_{j_o,\ldots,j_q,1},\ \ldots,\right.$$
$$\left.{}^{\sigma}a^{q,1-q}_{j_o,\ldots,j_q},\ \ldots,\ (-1)^{q-1}\, {}^{\sigma}a^{q+1,-q}_{j_o,\ldots,j_q,p}\right].$$

In dimension 1:

${}^{\sigma}M(a)_i^{o,1} = (0,\ldots,0,\ {}^{\sigma}a_i^{i,o},\ 0,\ldots,0)$, ${}^{\sigma}M(a)_{ji}^{1,o}$ has $r-s$ entry given by

$$\begin{cases} \delta_{rs} & r \neq i,j \\ 1 & r = i,\ s = i \\ -{}^{\sigma}a_{is}^{1,o} & r = i,\ s \neq i \\ {}^{\sigma}a_{ji}^{1,o} & r = j,\ s = i \\ {}^{\sigma}a_j^{o,1}\,{}^{\sigma}a_{jis}^{2,-1} & r = j,\ s \neq i \end{cases}$$

if $i \neq j$, and ${}^{\sigma}M(a)_{ii}^{1,o} = 1$, ${}^{\sigma}M(a)^{q,1-q} = 0$ for $q \geqslant 2$.

LEMMA. $(M(E)^{\cdot}, M(a))$ *is a simplicial twisting cochain that satisfies* (*1) *and* (*2).

Proof. A quick look at the formulas shows that (*1) and (*2) are satisfied, as are STC 1, 3, 4. Checking STC 2, namely the identity $\delta^{\sigma}M(a) + {}^{\sigma}M(a) \cdot {}^{\sigma}M(a) = 0$ is a six page matrix calculation that we do not reproduce. More specifically, Green checks [5]:

1) ${}^{\sigma}M(a)^{1,o}$ is a chain map in dimensions 0,1;

2) In dimension 1, ${}^{\sigma}M(a)_{kj}^{1,o}\, {}^{\sigma}M(a)_{ji}^{1,o} = {}^{\sigma}M(a)_{ki}^{1,o}$;

3) In dimension 0,

$$ {}^{\sigma}M(a)^{1,0}_{kj}\, {}^{\sigma}M(a)^{1,0}_{ji} + {}^{\sigma}M(a)^{0,1}_{k}\, {}^{\sigma}M(a)^{2,-1}_{kji} = {}^{\sigma}M(a)^{1,0}_{ki} \ ; $$

4) In dimension 0, the component of $\delta^{\sigma}M(a) + {}^{\sigma}M(a)\cdot{}^{\sigma}M(a)$ in bidegree $(q,2-q)$ is zero for $q > 2$.

Each of these is a straightforward but lengthy verification. Once these are verified, the rest of the twisting cochain identity is automatic, and STC 2 is proved.

Green's theorem now follows by induction, since the resolutions $M(E)^{\cdot}$ have the form

$$ \cdots \to M(E)^{-n+r+1}_{\sigma,\alpha} \to \cdots \to M(E)^{0}_{\sigma,\alpha} \to \cdots M(E)^{r+1}_{\sigma,\alpha} \to 0 \to \cdots , $$

and *1, *2 hold, thus $(M(E)^{\cdot}, M(a))$ is one step closer to a complex of simplicial bundles than (E,a) is. Namely, start with a twisting cochain for $\mathcal{S}$, apply the process M $n+1$ - times, and the result is a complex of simplicial bundles (n applications make $a^{q,1-q} = 0$ for $q > 1$, one more application makes the $E^{0}_{\sigma,\alpha}$ independent of α).

Relabel the last modification so that it is negatively graded, and call it $(E^{\cdot},a)$. This is Green's simplicial resolution.

§ 4. THE RIEMANN - ROCH THEOREM.

We now explain how Green's constructions can be combined with our proof of the Hirzebruch - Riemann - Roch formula for vector bundles [12] to extend this formula to coherent sheaf. Recall that the desired formula states that for a coherent sheaf $\mathcal{S}$ over a compact complex manifold M (or M open and $\mathcal{S}$ with compact support)

(HRR) $\chi(M,\mathcal{S}) = \int_M \mathrm{Todd}\,(M)\; \mathrm{ch}(\mathcal{S})$, where $\chi(M,\mathcal{S}) = \Sigma(-1)^i \dim H^i(M,\mathcal{S})$.

We start, as in [8,12], with the natural extension to the sheaf -theoretic context of Lefschetz's computation of the homology class of the diagonal in $M \times M$ [6], namely the commutative diagram

$$
\begin{array}{ccc}
\mathrm{Ext}^{n-i}(M,\mathcal{S},\Omega^n) \otimes H^i(M,\mathcal{S}) & \xrightarrow{tr} & H^n(M,\Omega^n) \xrightarrow{\int} \mathbb{C} \\
\downarrow \approx & & \uparrow tr \\
\mathrm{Ext}^n(M\times M,\ \pi_2^*\mathcal{S},\ \pi_1^*\mathcal{S}\otimes\Omega^{0,n}) & \xrightarrow{\Delta^*} & \mathrm{Ext}^n(M,\mathcal{S},\mathcal{S}\otimes\Omega^n) \\
\uparrow & & \\
\mathrm{Ext}^n(M\times M,\ \Delta_*\mathcal{S},\ \pi_1^*\mathcal{S}\otimes\Omega^{0,n}) & \xrightarrow{\approx} & H^0(M,\ \mathrm{Hom}(\mathcal{S},\mathcal{S})) \\
\Cup & & \Cup \\
\Lambda & \longleftarrow & \mathrm{id}
\end{array}
$$

For more details see [8]. We just recall that the pre-image of the identity under the lower horizontal arrow, which we denote by Λ for *Lefschetz class*, or dual class of the diagonal, can be obtained from a local computation. Comparison of the two ways of obtaining a number from Λ, corresponding to the two paths from the lower left hand corner to the upper right hand corner, gives HRR. Namely, the upper path gives quite formally $\chi(M,\mathcal{S})$, and the local nature of the situation forces the lower path to give Todd (M) ch($\mathcal{S}$).

In order to carry out the local computations we need a suitable complex to compute Ext. First, to compute Ext($M,\mathcal{S},\mathcal{S}$), rather than using the twisted complexes as in [8] we need to use a complex that is naturally suggested by Green's contruction, which we now describe.

Let $(E^{\cdot},\mathfrak{b})$ be a resolution of $\mathcal{S}$ by simplicial vector bundles. To avoid certain technicalities assume an ordering on $\mathfrak{U}$ as at the end of §2. By a k-*cochain* c *of compatible endomorphisms* of degree 1 we mean:

(C 1) For each $<\alpha_0\ldots\alpha_k>$, a collection

$$c_{\alpha_0\ldots\alpha_k} = \{{}^{\sigma}c_{\alpha_0\ldots\alpha_k}\}\ \sigma \geqslant <\alpha_0\ldots\alpha_k>$$

where ${}^{\sigma}c_{\alpha_0\ldots\alpha_k} : E^{\cdot}_{\sigma} \to E^{\cdot+1}_{\sigma}$.

(C2) Whenever $<\alpha_0\ldots\alpha_k> \leqslant \sigma < \tau$,

$${}^{\tau}c_{\alpha_0\ldots\alpha_k} = \begin{bmatrix} {}^{\sigma}c_{\alpha_0\ldots\alpha_k} & * \\ 0 & 0 \end{bmatrix} .$$

Denote the collection of all such cochains by $C^k(\mathfrak{U}, \mathrm{Hom}^1(E,E))$. Define a Cech coboundary $\delta: C^k(\mathfrak{U}, \mathrm{Hom}^1(E,E)) \to C^{k+1}(\mathfrak{U}, \mathrm{Hom}^1(E,E))$ by letting $(\delta c)_{\alpha_0\ldots\alpha_{k+1}}$ be the $k+1$ - cochain

$$\left\{ \sum_{i=0}^{k} (-1)^i\, {}^{\sigma}c_{\alpha_0\ldots\hat{\alpha}_i\ldots\alpha_{k+1}} + (-1)^{k+1}\, {}^{\sigma}b^{-1}_{\alpha_k\alpha_{k+1}}\, {}^{\sigma}c_{\alpha_0\ldots\alpha_k}\, {}^{\sigma}b_{\alpha_k\alpha_{k+1}} \right\} ,$$

for all $\sigma \geqslant <\alpha_0\ldots\alpha_{k+1}>$.

Define a differential raising 1 degree by one by the usual formula for Hom of two complexes. The two differentials commute, and the resulting bicomplex $C^{\cdot}(\mathfrak{U}, \mathrm{Hom}^{\cdot}(E,E))$ is called the *complex of cochains of compatible endomorphisms*. Standard arguments as in [7,13] show that the cohomology of the total complex is Ext($M,\mathcal{S},\mathcal{S}$). This complex has a naturally defined trace

$$C^{\cdot}(\mathfrak{U}, \mathrm{Hom}^{\cdot}(E,E)) \to C^{\cdot}(\mathfrak{U},\mathcal{O})$$

$$c_{\alpha_0\ldots\alpha_k} \to \mathrm{tr}\left[{}^{<\alpha_0\ldots\alpha_k>}c_{\alpha_0\ldots\alpha_k}\right],$$

where tr means Lefschetz number (alternating sum of trace) for endomorphisms of degree zero, and zero otherwise. This is easily verified to be a chain map (just as in the Chech - de Rham identity of §2). Thus we get

$$\mathrm{Ext}^{\cdot}(M,\mathcal{S},\mathcal{S}) \xrightarrow{\mathrm{tr}} H^{\cdot}(M,\mathcal{O}).$$

Tensoring the construction with Ω^n, we get

$$C^{\cdot}(\mathfrak{U}, \mathrm{Hom}^{\cdot}(E,E\otimes\Omega^n)) \xrightarrow{\mathrm{tr}} C^{\cdot}(\mathfrak{U},\Omega^n)$$

which induces the map

$$\mathrm{Ext}^n(M,\mathcal{S},\mathcal{S}\otimes\Omega^n) \xrightarrow{\mathrm{tr}} H^n(M, \Omega^n)$$

in the top right hand part fo the diagram.

The group $\mathrm{Ext}^n(M\times M, \pi_2^*\mathcal{S}, \pi_1^*\mathcal{S}\otimes\Omega^{0,n})$ is computed by a similar complex of cochains of compatible homomorphisms. The group $\mathrm{Ext}^n(M\times M, \Delta_*\mathcal{S}, \pi_1^*\mathcal{S}\otimes\Omega^{0,n})$, in which Λ lies, is computed by tensoring the dual twisted complex of the Koszul resolutions of $\mathcal{O}_\Delta$ with $\mathrm{Hom}^{\cdot}(\pi_2^*E, \pi_1^*E\otimes\Omega^{0,n})$. The point is that the structure of the simplicial resolution $(E^{\cdot},b)$ is sufficiently close to that of a vector bundle that the formula given in [12], §12 for tensoring the twisting cochain with $\mathrm{Hom}(\pi_2^*E, \pi_1^*E)\otimes\Omega^{0,n}$ (E a vector bundle) applies, with the obvious interpretation, to $(E^{\cdot},b)$. We explain this only over the subcomplex of the nerve of $\mathfrak{U}\times\mathfrak{U}$ spanned by the vertices $U_\alpha\times U_\alpha$, since the restriction of Λ to the diagonal factors through this.

A k-cochain with values in $\check{K}\otimes\mathrm{Hom}^{\cdot}(\pi_2^*E, \pi_1^*E)\otimes\Omega^{0,n} = \mathrm{Hom}^{\cdot}(\pi_1^*E, \pi_2^*E\otimes\check{K}\otimes\Omega^{0,n})$ is a collection

$$c_{\alpha_0\ldots\alpha_k} = \left\{{}^{\sigma}c_{\alpha_0\ldots\alpha_k} : \pi_2^*E_\sigma^{\cdot} \to \pi_1^*E_\sigma^{\cdot}\otimes\check{K}\otimes\Omega^{0,n}\right\}_{\sigma\geq<\alpha_0\ldots\alpha_k>}$$

which satisfies the compatibility condition C 2 with respect to the bases e_σ, e_τ and e_σ, e_τ tensored with standard basis for $\check{K}\otimes\Omega^{0,n}$. The twisting cochain $\check{a}$ for $\check{K}$ acts by letting $(\check{a}^{1+1,-1}\cdot c)_{\alpha_0\ldots\alpha_{k+1}}$ be $\pm$ the collection

$$\left\{\check{a}_{\alpha_k\cdots\alpha_{k+1}}\ \pi_1^*\ {}^{\sigma}b^{-1}_{\alpha_k\alpha_{k+1}}\ {}^{\sigma}c_{\alpha_0\cdots\alpha_k}\ \pi_2^*\ {}^{\sigma}b_{\alpha_k\alpha_{k+1}}\right\}_{\sigma\geqslant\langle\alpha_0\cdots\alpha_{k+1}\rangle}\ .$$

The general argument of [8],§2 can be refeated in this context to construct Λ. Examination of the formulas shows that the final expression for $\chi(M,\mathcal{S})$ is the same as applying the process of [12], §12 to each constituent simplicial bundle (E',b) of $(E^{\cdot},b)$ and taking alternating sum, hence by the invariance argument these (or the more direct analysis of [8], §4) the answer is Todd (M) $ch^A(\mathcal{S})$. But by the Hodge filtration of $ch^G(\mathcal{S})$ (§2), this has the same evaluation over M as Todd (M) ch $(\mathcal{S})$ (cf [12],§1), and HRR is proved.

We have ommitted many details since all that is really required is a careful reading of [12] and [8], §2, checking that everything goes though in this context. For the reader interested in following the details, we mention two facts that are needed:

1) The "initial condition" for constructing Λ has a 1 in the lower right hand corner of C 2, rather than a zero, although all the other terms in the successive construction have a zero. It is easy to see that the complex obtained by putting an arbitrary entry in that corner is quasi-isomorphic to the one we defined.

2) The expression for Λ obtained from [8], §2 looks quite complicated, since it also involves the resolution differential, but all these terms have positive endomorphism degree, hence zero trace. Thus only the transiton functions (i.e., the expressions in [12],§12) appear in the final answer.

BIBLIOGRAPHY

[1] M.F. Atiyah. *Complex analytic onnections in fibre bundles.* Trans.Am.Math. Soc. 85(1957), 181-207.

[2] M.F. Atiyah and F. Hirzebruch. *Analytic cycles on complex manifolds.* Topology 1(1962), 25-45.

[3] M.F. Atiyah and F. Hirzebruch. *The Riemann-Roch theorem for analytic embeddings.* Topology 1(1962), 151-166.

[4] R . Bott. *Lectures on characteristic classes and foliations.* In lectures on Algebraic and Differential Topology, Lecture Notes in Mathematics 279, Springer Verlag, Berlin-Heidelberg, New York, 1972.

[5] H.I. Green. *Chern classes for coherent sheaves.* Ph.D. Thesis, University of Warwick, January 1980.

[6] S. Lefschetz. *Intersections and transformations of complexes and manifolds.* Trans.Am.Math.Soc. 28(1926), 1-49.

[7] N.R. O'Brian, D. Toledo and Y.L.L. Tong. *A trace map and characteristic classes for coherent sheaves.* Amer. J. Math. 103(1981), 225-252.

[8] N.R. O'Brian, D. Toledo and Y.L.L. Tong. *Hirzebruch - Riemann - Roch for coherent sheaves*. Amer. J. Math. 103(1981), 253-271.

[9] N.R. O'Brian, D. Toledo and Y.L.L. Tong. *Grothendieck - Riemann - Roch for analytic maps of complex manifolds*. Math.Ann. (1985).

[10] H.W. Schuster. *Locally free resolutions of coherent sheaves on surfaces*. J. Reine Ang W. Math. 337(1982), 159-165.

[11] I.M. Singer. *Future extensions of index theory and elliptic operators*. Annals.Math. Studies 70(1971).

[12] D. Toledo and Y.L.L. Tong. *A parametrix for* $\bar{\partial}$ *and Riemann - Roch in Cech theory*. Topology 15(1976), 273-301.

[13] D. Toledo and Y.L.L. Tong. *Duality and intersection theory in complex manifolds, I*. Math.Ann. 237(1978), 41-77.

DEPARTMENT OF MATHEMATICS
University of UTAH
Salt Lake City, UTAH 84112, U.S.A.

DEPARTMENT OF MATHEMATICS
Purdue University
West Lafayette, Indiana 47907,U.S.A.

ABCDEFGHIJ - 89876